Genetics Simplified: A Comprehensive Guide for NEET Examination"

By: Mir Hilal

Overview:

"Genetics Simplified: A Comprehensive Guide for NEET Examination" is an essential E-book designed to help students preparing for the NEET (National Eligibility cum Entrance Test) examination gain a thorough understanding of genetics. This comprehensive guide provides a simplified approach to the complex concepts of genetics, ensuring that students grasp the fundamentals with ease.

In this e-book, you will find a well-structured and organized presentation of genetic principles, making it easier for students to navigate through the subject. The content is tailored specifically to meet the requirements of the NEET examination, ensuring that students focus on the topics that are most likely to appear in the test.

Key features of "Genetics Simplified" include:

1. Comprehensive Coverage: The e-book covers all the important topics related to genetics, including the structure and function of DNA, gene expression and regulation, Mendelian genetics, population genetics, genetic disorders, and much more. Each topic is explained in a concise and easy-to-understand manner.
2. Simplified Explanations: Complex genetic concepts are broken down into simplified explanations, ensuring that even students with minimal prior knowledge of genetics can grasp the content. The book uses clear language and avoids unnecessary jargon, making it accessible to a wide range of readers.
3. Visual Aids: The e-book is enriched with diagrams, illustrations, and tables to enhance the understanding of genetic concepts.

These visual aids provide a visual representation of the concepts, making them easier to comprehend and remember.

4. Practice Questions: To reinforce learning and help students assess their understanding, "Genetics Simplified" includes a wide range of practice questions with detailed explanations. These questions are designed to simulate the NEET examination format and cover various difficulty levels.

5. Tips and Strategies: The e-book offers valuable tips and strategies to help students tackle genetic-based questions effectively. It provides guidance on how to approach different types of questions, identify key information, and apply genetic principles to solve problems.

"Genetics Simplified: A Comprehensive Guide for NEET Examination" is a valuable resource for NEET aspirants looking to build a strong foundation in genetics. Whether you are a beginner or seeking to strengthen your knowledge in genetics, this e-book will provide you with the necessary tools and insights to excel in the NEET examination and pursue a career in the medical field.

Contents:

Chapter 1: Introduction to Genetics

- Historical perspective of genetics
- The structure and function of DNA
- Genes and chromosomes
- Central dogma of molecular biology

Chapter 2: Mendelian Genetics

- Mendel's laws of inheritance
- Monohybrid and dihybrid crosses
- Punnett squares and probability
- Deviations from Mendelian inheritance (incomplete dominance, codominance, multiple alleles)

Chapter 3: Extensions of Mendelian Genetics

- Sex-linked inheritance
- Pedigree analysis
- Genetic disorders (autosomal and sex-linked)
- Genetic counseling and prenatal testing

Chapter 4: Molecular Genetics

- DNA replication, transcription, and translation
- Genetic code and protein synthesis
- Gene regulation and control
- Mutations and their consequences

Chapter 5: Population Genetics

- Hardy-Weinberg equilibrium
- Factors affecting gene frequency (selection, genetic drift, gene flow)
- Genetic variation and human populations
- Speciation and evolutionary processes

Chapter 6: Chromosomal Basis of Inheritance

- Chromosome structure and organization
- Sex determination and sex chromosomes
- Chromosomal aberrations (aneuploidy, polyploidy)
- Linkage and recombination

Chapter 7: Biotechnology and Genetic Engineering

- Recombinant DNA technology
- DNA cloning and genetic transformation
- Genetically modified organisms (GMOs)
- Applications in medicine, agriculture, and industry

Chapter 8: Genomics and Personalized Medicine

- Human genome project and sequencing technologies
- Comparative genomics and functional genomics
- Pharmacogenomics and precision medicine
- Ethical considerations in genomics research

Chapter 9: Gene Expression and Regulation

- Transcription factors and gene regulation networks
- Epigenetics and chromatin remodeling
- Non-coding RNAs (microRNAs, long non-coding RNAs)
- Gene expression profiling techniques

Chapter 10: Genetic Techniques and Tools

- Polymerase chain reaction (PCR)
- DNA sequencing methods
- Gel electrophoresis and DNA fingerprinting
- Bioinformatics and computational biology

Chapter 11: Applied Genetics

- Medical genetics and genetic disorders
- Forensic genetics and DNA profiling
- Agricultural biotechnology and crop improvement
- Conservation genetics and biodiversity preservation

Chapter 12: Practice Questions and Mock Tests

- NEET-style questions and explanations
- Mock tests to assess understanding and preparedness

Appendix: Glossary of Genetic Terms

Chapter 1:
Introduction to Genetics

· **Historical perspective of genetics**
· **The structure and function of DNA**
· **Genes and chromosomes**
· **Central dogma of molecular biology**

Introduction:

Genetics is the branch of biology that studies how traits are passed from one generation to the next. It encompasses the study of genes, heredity, and variation in living organisms. The field of genetics provides insights into the fundamental principles of life, including the mechanisms that govern inheritance, the diversity of organisms, and the basis for genetic diseases.

At the core of genetics is the concept of genes, which are segments of DNA (deoxyribonucleic acid) that contain the instructions for building and maintaining an organism. DNA, found in the nucleus of cells, carries the genetic information that determines an organism's characteristics, such as its physical features, behaviors, and susceptibility to certain diseases.

Genes exist in pairs called alleles, with one allele inherited from each parent. These alleles can be either dominant or recessive, meaning that one allele may have a stronger effect on the trait it controls than the other. The specific combination of alleles an individual possesses determines their genotype, while the observable traits resulting from these alleles make up their phenotype.

The study of genetics involves exploring various concepts and mechanisms, such as:

1. **Mendelian Inheritance:** Gregor Mendel, an Austrian monk, laid the foundation for the modern study of genetics in the 19th century. He discovered patterns of inheritance by

conducting experiments on pea plants. Mendel's laws, known as the laws of inheritance, describe how traits are passed from parents to offspring.

2. **Chromosomes and Genomes:** Genes are organized into structures called chromosomes. In humans, for example, chromosomes are found in the nucleus of each cell and occur in pairs, with one copy inherited from each parent. The complete set of genetic information within an organism is called its genome.

3. **DNA Structure and Replication:** DNA is a double-stranded molecule formed by a sequence of nucleotides. Each nucleotide consists of a sugar, a phosphate group, and one of four nitrogenous bases: adenine (A), thymine (T), cytosine (C), and guanine (G). The specific order of these bases forms the genetic code. DNA replication is the process by which a cell duplicates its DNA before cell division, ensuring that genetic information is faithfully transmitted to daughter cells.

4. **Genetic Variation:** Genetic variation refers to the diversity of genetic information within a population or species. It arises from different combinations of alleles, mutations, and recombination during sexual reproduction. Genetic variation is essential for adaptation and evolution.

5. **Genetic Disorders and Genetic Counseling:** Certain genetic variations can lead to genetic disorders, which are conditions caused by abnormal genes or chromosomal changes. Genetic counseling is a specialized field that helps individuals and families understand their risk of inherited disorders, make informed reproductive decisions, and manage the impact of genetic conditions.

6. **Molecular Genetics:** Molecular genetics focuses on the study of genes at the molecular level, including their structure, function, and regulation. It involves techniques such as DNA sequencing, polymerase chain reaction (PCR), and gene expression analysis to unravel the complexities of genetic information.

The field of genetics has numerous applications in various areas, including medicine, agriculture, forensics, and biotechnology. It has revolutionized our understanding of heredity, evolution, and the molecular basis of life, paving the way for advancements in personalized medicine, genetic engineering, and other areas of scientific research.

Historical perspective of genetics:

The field of genetics has a rich and fascinating history that spans several centuries. Here is a brief historical perspective on genetics:

1. **Gregor Mendel and the Birth of Genetics (1860s):** Gregor Mendel, an Austrian monk, is often referred to as the "Father of Genetics." In the 1860s, he conducted groundbreaking experiments on pea plants, carefully studying their inheritance patterns. Through his meticulous work, Mendel formulated the laws of inheritance, known as Mendelian genetics. He proposed the concepts of dominant and recessive traits, as well as the segregation and independent assortment of genes.

2. **Rediscovery of Mendel's Work (1900):** Mendel's work went largely unnoticed until the early 20th century when it was independently rediscovered by three scientists: Carl Correns, Hugo de Vries, and Erich von Tschermak. They recognized the significance of Mendel's experiments and his laws of inheritance, laying the foundation for modern genetics.

3. **Chromosome Theory of Inheritance (early 20th century):** In the early 1900s, researchers began to study the connection between Mendelian inheritance and the physical structures within cells. Thomas Hunt Morgan and his colleagues conducted extensive experiments on fruit flies and proposed the chromosome theory of inheritance. They discovered that genes are located on chromosomes and that the behavior of chromosomes during cell division explains the patterns of inheritance.

4. **Discovery of DNA as the Genetic Material (1940s-1950s):** Prior to the mid-20th century, the nature of the genetic material remained unknown. However, through a series of

experiments, scientists established that DNA (deoxyribonucleic acid) is the molecule responsible for carrying genetic information. Key contributions came from researchers such as Oswald Avery, Colin MacLeod, and Maclyn McCarty, who demonstrated that DNA, not proteins, was the transforming principle in bacteria. The groundbreaking discovery of the DNA double helix structure by James Watson and Francis Crick in 1953 further solidified DNA's role as the genetic material.

5. **Advances in Molecular Genetics** (1960s onwards): The field of genetics witnessed rapid advancements in the second half of the 20th century. Techniques like polymerase chain reaction (PCR), developed by Kary Mullis in the 1980s, enabled the amplification of specific DNA sequences. The advent of DNA sequencing methods, starting with Sanger sequencing in the 1970s, paved the way for deciphering the entire human genome, which was accomplished with the Human Genome Project in 2003.

6. **Modern Genomics and Genetic Engineering:** The completion of the Human Genome Project marked a turning point in genetics. It opened the doors to studying entire genomes and exploring the role of genes in health, disease, and evolution. Advances in genomics, such as next-generation sequencing technologies, have made it possible to sequence genomes quickly and cost-effectively. Genetic engineering techniques like CRISPR-Cas9 have revolutionized gene editing, allowing precise modifications of DNA sequences.

Today, genetics plays a critical role in various fields, including medicine, agriculture, forensics, and evolutionary biology. The ongoing advancements in genetics continue to deepen our understanding of life's complexity and hold promise for solving numerous biological challenges.

The structure and function of DNA:

DNA, short for deoxyribonucleic acid, is a molecule that carries the genetic instructions for the development, functioning, growth, and reproduction of all known living organisms. It serves as the blueprint for building and maintaining an organism. Let's delve into the structure and function of DNA.

Structure of DNA: DNA has a double helix structure, resembling a twisted ladder or a spiral staircase. It consists of two long strands made up of nucleotides. Each nucleotide is composed of three components:

1. **Deoxyribose sugar:** It is a five-carbon sugar molecule that forms the backbone of the DNA strand.
2. **Phosphate group:** It is a phosphorus atom bonded to four oxygen atoms, providing a negative charge.
3. **Nitrogenous base:** There are four types of nitrogenous bases in DNA: adenine (A), thymine (T), cytosine (C), and guanine (G). The bases pair up in a specific manner: A always pairs with T, and C always pairs with G. These pairs are held together by hydrogen bonds.

The DNA strands run antiparallel to each other, meaning they align in opposite directions. One strand runs in the 5' to 3' direction, while the other runs in the 3' to 5' direction. The complementary base pairing ensures that the genetic information is accurately copied during DNA replication.

Function of DNA:

1. **Storage of genetic information:** DNA carries the genetic instructions required for the development, functioning, and traits of an organism. It contains the complete set of genes, which are segments of DNA that encode specific proteins or regulatory molecules.
2. **Replication:** DNA undergoes a process called DNA replication, where it duplicates itself before cell division. This ensures that each daughter cell receives an identical copy of the genetic information. During replication, the DNA strands separate, and each strand serves as a template for the synthesis of a new complementary strand. This process ensures genetic continuity and inheritance.
3. **Gene expression:** DNA contains genes that serve as templates for the synthesis of proteins, which are the

building blocks of cells and perform various functions in the body. The process of gene expression involves two stages: transcription and translation. Transcription occurs in the cell nucleus, where the DNA sequence of a gene is transcribed into a molecule called messenger RNA (mRNA). The mRNA then moves to the cytoplasm, where it is translated into a specific protein by ribosomes.

4. **Genetic variation and evolution:** DNA is subject to changes called mutations, which are alterations in the DNA sequence. Mutations can occur spontaneously or be induced by external factors such as radiation or chemicals. These mutations, when inherited or occurring in reproductive cells, can lead to genetic variation within a population. Genetic variation is the raw material for evolution and contributes to the diversity of species over time.

In summary, DNA's structure as a double helix and its function as the carrier of genetic information are essential for the growth, development, reproduction, and functioning of all living organisms.

Genes and chromosomes:

Genes and chromosomes are fundamental components of living organisms, including humans. They play a crucial role in determining an organism's inherited traits, genetic information, and overall development. Let's explore these concepts in more detail:

1. **Genes:**
    - Genes are segments of DNA (deoxyribonucleic acid) that contain instructions for building and maintaining an organism.
    - They are the basic units of heredity and are responsible for transmitting traits from parents to offspring.
    - Genes encode specific information that guides the synthesis of proteins, which are essential for various biological functions.
    - Each gene is located at a specific position on a chromosome.
2. **Chromosomes:**

- Chromosomes are structures within cells that contain DNA.
- They are thread-like structures made up of tightly coiled DNA and associated proteins.
- Chromosomes are found in the nucleus of eukaryotic cells, including human cells.
- Humans typically have 46 chromosomes in their cells, organized into 23 pairs. These pairs are called homologous chromosomes.
- In each pair, one chromosome is inherited from the mother, and the other is inherited from the father.
- The first 22 pairs of chromosomes are called autosomes, and the last pair determines an individual's sex (XX for females and XY for males).

3. **Relationship between Genes and Chromosomes:**
- Genes are located on chromosomes and exist as specific sequences of DNA.
- Each chromosome carries many genes, and the exact number can vary depending on the organism.
- Humans have approximately 20,000 to 25,000 genes distributed across their 46 chromosomes.
- The position of a gene on a chromosome is called a locus.
- During cell division, chromosomes are replicated, and each daughter cell receives a complete set of chromosomes, including the genes they carry.

4. **Gene Expression:**
- Gene expression refers to the process by which the information stored in a gene is used to create a functional product, usually a protein.
- Gene expression is tightly regulated and can be influenced by various factors, including environmental conditions and internal cellular signals.

- Not all genes are actively expressed at all times; different cells and tissues have unique patterns of gene expression, allowing for specialized functions.

Understanding genes and chromosomes is crucial for comprehending the mechanisms of inheritance, genetic disorders, and the diversity of traits observed in living organisms. Advances in genetics and genomics have greatly expanded our knowledge of these topics, contributing to various fields such as medicine, agriculture, and evolutionary biology.

Central dogma of molecular biology:

The central dogma of molecular biology is a fundamental principle that describes the flow of genetic information within a biological system, particularly in cells. It explains how genetic information is stored, replicated, transcribed into RNA, and translated into proteins. The central dogma can be summarized in three main steps:

1. **DNA Replication:** The process by which DNA molecules are duplicated. During replication, the two strands of the DNA molecule separate, and each strand serves as a template for the synthesis of a new complementary strand. This results in two identical DNA molecules, each containing one original and one newly synthesized strand.

2. **Transcription:** The process by which genetic information in DNA is copied into RNA molecules. In transcription, an enzyme called RNA polymerase binds to a specific region of DNA called a promoter and synthesizes a complementary RNA molecule based on the DNA template. This RNA molecule is called messenger RNA (mRNA) and carries the genetic information from the DNA to the next step.

3. **Translation:** The process by which the genetic information carried by mRNA is used to synthesize proteins. Translation occurs in the ribosomes, where transfer RNA (tRNA) molecules, with their specific anticodons, bind to the mRNA codons and bring the corresponding amino acids. The ribosome then catalyzes the formation of peptide bonds

between the amino acids, resulting in the synthesis of a polypeptide chain. This chain folds into a functional protein. In summary, the central dogma states that genetic information flows from DNA to RNA through transcription and then from RNA to protein through translation. It underlies the fundamental processes of gene expression and protein synthesis in living organisms.

Chapter 2:
Mendelian Genetics

- Mendel's laws of inheritance
- Monohybrid and dihybrid crosses
- Punnett squares and probability
- Deviations from Mendelian inheritance (incomplete dominance, codominance, multiple alleles)

Introduction:

Mendelian genetics, named after the Austrian monk Gregor Mendel, is a branch of genetics that focuses on the study of inheritance in organisms. Mendel's groundbreaking work in the mid-19th century laid the foundation for modern genetics. His experiments with pea plants led to the formulation of several fundamental principles of inheritance, which are now known as Mendel's laws.

Mendelian genetics revolves around the concept of genes, which are segments of DNA that contain instructions for the development and functioning of living organisms. Genes are passed down from parents to offspring and determine various traits, such as eye color, height, and susceptibility to certain diseases.

Mendel's laws of inheritance:

Gregor Mendel formulated three laws of inheritance based on his experiments with pea plants. These laws are the foundation of modern genetics and are still relevant today. Here are Mendel's three laws of inheritance:

1. **Law of Segregation:** According to this law, during the formation of gametes (reproductive cells), the two alleles (alternative forms of a gene) for a trait segregate or separate from each other, so that each gamete carries only one allele for that trait. This means that offspring receive one allele from each parent, and the alleles segregate equally into the gametes.

For example, if a pea plant has a pair of alleles for flower color, one for purple flowers (P) and one for white flowers (p), the law of segregation states that during gamete formation, half of the gametes will carry the allele for purple flowers (P), and the other half will carry the allele for white flowers (p).

2. **Law of Independent Assortment:** According to this law, the segregation of alleles for one trait is independent of the segregation of alleles for another trait. In other words, the inheritance of one trait does not influence the inheritance of another trait unless the genes for both traits are located very close to each other on the same chromosome.

For example, if Mendel studied two traits in his pea plants, such as flower color and seed shape, the law of independent assortment states that the alleles for flower color (purple or white) will segregate independently of the alleles for seed shape (round or wrinkled). This means that all possible combinations of flower color and seed shape are equally likely in the offspring.

3. **Law of Dominance:** According to this law, when two different alleles for a trait are present, one allele (the dominant allele) will be expressed phenotypically, while the other allele (the recessive allele) remains unexpressed unless two copies of it are present. The dominant allele masks the presence of the recessive allele in the phenotype of an individual.

For example, if a pea plant has one allele for purple flowers (dominant, P) and one allele for white flowers (recessive, p), the dominant allele (P) will determine the flower color, and the plant will have purple flowers. The recessive allele (p) will only be expressed if both alleles are recessive (pp), resulting in white flowers.

These laws of inheritance provided the framework for understanding the transmission of traits from parents to offspring and helped establish the principles of genetic inheritance.

Monohybrid and dihybrid crosses:

Monohybrid and dihybrid crosses are two types of genetic crosses used in the field of genetics to study the inheritance of traits between organisms. These crosses involve the mating of individuals that differ in one or two traits, respectively. Let's explore each type of cross in more detail:

1. **Monohybrid Cross:**
    - Definition: A monohybrid cross involves the mating of two individuals that differ in only one trait.
    - Trait: The cross focuses on the inheritance of a single trait, which is typically controlled by a single gene with two different alleles.
    - Alleles: Each parent carries two alleles for the trait, one inherited from each parent. The alleles can be either dominant or recessive.
    - Offspring: The resulting offspring, known as the F1 generation, will inherit one allele from each parent. Depending on the dominance or recessiveness of the alleles, the offspring's phenotype may display the dominant trait, the recessive trait, or a combination of both.
    - Example: A classic example of a monohybrid cross is the mating between two pea plants that differ in flower color, one with yellow flowers (YY) and the other with green flowers (yy). The resulting offspring will all have yellow flowers in the F1 generation, as the yellow allele (Y) is dominant over the green allele (y).
2. **Dihybrid Cross:**
    - Definition: A dihybrid cross involves the mating of two individuals that differ in two traits.

15

- Traits: The cross focuses on the inheritance of two different traits, each controlled by a separate gene with two different alleles.
- Alleles: Each parent carries two alleles for each trait, and these alleles can be dominant or recessive.
- Offspring: The resulting offspring, known as the F1 generation, will inherit one allele for each trait from each parent. The combinations of alleles will determine the phenotype of the offspring.
- Mendel's Law of Independent Assortment: The dihybrid cross also allows us to observe Mendel's Law of Independent Assortment, which states that genes for different traits segregate independently during the formation of gametes.
- Example: An example of a dihybrid cross is the mating between pea plants that differ in seed color (yellow vs. green) and seed texture (smooth vs. wrinkled). The alleles for seed color are represented by Y (yellow) and y (green), while the alleles for seed texture are represented by S (smooth) and s (wrinkled). The cross could involve a plant with genotype YYSS and another with genotype yyss. The F1 generation will all have yellow and smooth seeds since the dominant alleles (Y and S) mask the recessive alleles (y and s).

Both monohybrid and dihybrid crosses provide insights into the patterns of inheritance of specific traits and help scientists understand how genes are passed down from one generation to the next. These crosses are fundamental tools in genetics research and have contributed significantly to our understanding of genetics.

Punnett squares and probability:

Punnett squares are a useful tool in genetics to predict the probability of certain traits being inherited in offspring. They are named after Reginald Punnett, a British geneticist who developed this method in the early 20th century.

A Punnett square is a visual representation of the possible combinations of alleles from two parents. Alleles are alternative forms of a gene, and they determine the traits or characteristics that an organism inherits. In a Punnett square, the alleles of one parent are written along the top of the square, and the alleles of the other parent are written along the side of the square.

To calculate the probabilities of different outcomes, you need to understand a few key concepts:

1. **Dominant and recessive alleles:** An allele can be dominant or recessive. Dominant alleles are expressed in the phenotype (the physical appearance or trait) even if only one copy is present, while recessive alleles are only expressed if two copies are present.

2. **Homozygous and heterozygous:** Homozygous individuals have two copies of the same allele for a particular gene (e.g., AA or aa), while heterozygous individuals have two different alleles (e.g., Aa).

3. **Genotype and phenotype:** The genotype refers to the combination of alleles an organism possesses, while the phenotype refers to the physical expression of those alleles.

To use a Punnett square, follow these steps:

1. Determine the genotypes of the parents. For example, if one parent is homozygous dominant (AA) and the other is homozygous recessive (aa), write these genotypes along the top and side of the square, respectively.

2. Fill in the boxes of the Punnett square by combining the alleles from each parent. For example, combine the A allele from the first parent with the a allele from the second parent. This will give you all the possible combinations of alleles in the offspring.

3. Determine the phenotype for each genotype in the square. If the dominant allele (A) is present, it will determine the phenotype. If both alleles are recessive (aa), the recessive trait will be expressed.

4. Count the number of squares with each phenotype to determine the probability of inheriting a specific trait. Divide

the number of squares showing a particular trait by the total number of squares.

Remember, Punnett squares provide predictions based on the laws of probability and assume that alleles segregate independently. While they can give you an idea of the expected outcomes, actual genetic inheritance can be more complex due to factors like incomplete dominance, codominance, polygenic inheritance, and gene interactions.

It's important to note that Punnett squares are just a tool for calculating probabilities and do not guarantee the actual outcome in a real-life scenario. They are helpful for understanding the basic principles of inheritance and estimating the likelihood of certain traits appearing in offspring.

Deviations from Mendelian inheritance (incomplete dominance, codominance, multiple alleles):

Deviations from Mendelian inheritance refer to cases where the patterns of inheritance do not strictly follow Gregor Mendel's laws of inheritance. Mendel's laws describe the principles of dominant and recessive alleles, as well as the segregation and independent assortment of alleles. However, there are several situations in genetics where these patterns are not observed. Three common deviations from Mendelian inheritance are incomplete dominance, codominance, and multiple alleles.

1. **Incomplete Dominance:** Incomplete dominance occurs when the phenotype of a heterozygous individual is an intermediate blend of the phenotypes of the two homozygous individuals. Neither allele is completely dominant or recessive, resulting in a third phenotype. For example, in snapdragons, crossing a red-flowered plant (RR) with a white-flowered plant (WW) produces pink-flowered offspring (RW).

2. **Codominance:** Codominance occurs when both alleles in a heterozygous individual are fully expressed, without any blending. In other words, both alleles contribute to the phenotype simultaneously. A classic example is the human ABO blood group system. The A and B alleles are

codominant, so an individual with genotype AB will have both A and B antigens expressed on their red blood cells.

3. **Multiple Alleles:** Multiple alleles refer to a situation where a gene has more than two possible alleles within a population. However, an individual can only have a maximum of two alleles, one inherited from each parent. A well-known example is the ABO blood group system mentioned earlier. In addition to the A and B alleles, there is a third allele, O, which is recessive to both A and B. The combinations of these alleles determine the ABO blood types.

It's important to note that these deviations from Mendelian inheritance do not invalidate Mendel's principles but rather expand our understanding of genetic inheritance and account for the complexity of genetic traits observed in real-world scenarios.

Chapter 3:
Extensions of Mendelian Genetics

· **Sex-linked inheritance**
· **Pedigree analysis**
· **Genetic disorders (autosomal and sex-linked)**
· **Genetic counseling and prenatal testing**

Introduction:

Extensions of Mendelian genetics refer to the advancements and modifications made to Gregor Mendel's principles of inheritance, which formed the basis of classical or Mendelian genetics. While Mendelian genetics focuses on the patterns of inheritance of single genes, these extensions explore additional complexities and phenomena that arise in genetics.

1. **Incomplete Dominance and Codominance:** Mendel's laws describe dominant and recessive alleles, but in some cases, neither allele is fully dominant, resulting in incomplete dominance. This means that the heterozygous phenotype is

an intermediate blend of the two homozygous phenotypes. Codominance, on the other hand, occurs when both alleles in a heterozygote are expressed equally and distinctly.

2. **Multiple Alleles:** Mendel's experiments involved two alleles per gene, but many genes exist in multiple forms called alleles. Multiple alleles refer to the presence of more than two alternative forms of a gene within a population, creating a greater variety of genotypes and phenotypes.

3. **Polygenic Inheritance:** Mendelian genetics focuses on traits controlled by a single gene, but many traits are influenced by the combined effects of multiple genes. Polygenic inheritance refers to the inheritance of traits that are controlled by the cumulative effects of multiple genes, resulting in a wide range of phenotypic variations.

4. **Pleiotropy:** Mendel's principles assume that each gene controls a single trait, but some genes can have multiple effects on different traits or systems. This phenomenon is known as pleiotropy, where a single gene can influence multiple, seemingly unrelated phenotypic traits.

5. **Epistasis:** Epistasis occurs when the expression of one gene masks or modifies the expression of another gene. It involves the interaction of genes at different loci, affecting the phenotypic ratio observed in the offspring.

6. **Gene Interactions:** In addition to epistasis, there are other types of gene interactions that affect inheritance patterns. These include complementarity, where two genes work together to produce a phenotype, and gene suppression, where the presence of one gene suppresses the expression of another gene.

7. **Sex-Linked Inheritance:** Mendel's experiments did not consider the inheritance patterns of genes located on the sex chromosomes. Sex-linked inheritance refers to the inheritance of genes located on the sex chromosomes (usually the X chromosome in humans), leading to distinctive patterns of inheritance, such as X-linked recessive disorders.

8. **Environmental Influence:** Mendelian genetics primarily focuses on the role of genes in determining phenotypes. However, environmental factors can also influence gene expression and contribute to phenotypic variation. Gene-environment interactions play a significant role in the inheritance of complex traits.

These extensions to Mendelian genetics have provided a more comprehensive understanding of inheritance patterns, genetic variation, and the complexities of gene regulation. They have contributed to our knowledge of human genetics, evolution, and the understanding of various genetic disorders.

Sex-linked inheritance:

Sex-linked inheritance refers to the inheritance of genes located on the sex chromosomes, specifically the X and Y chromosomes. In humans, sex-linked inheritance patterns are commonly associated with genes on the X chromosome because it is larger and carries more genes compared to the Y chromosome.

The X and Y chromosomes determine an individual's biological sex. Females have two X chromosomes (XX), while males have one X and one Y chromosome (XY). Since males have only one copy of the X chromosome, any gene present on it will be expressed, regardless of whether it is dominant or recessive. This means that males are more likely to be affected by X-linked genetic disorders if they inherit a mutated gene on their X chromosome.

There are two main types of sex-linked inheritance:

1. X-linked recessive inheritance: In this pattern, a mutation in a gene on the X chromosome leads to the expression of a recessive trait or disorder. Males who inherit the mutated gene on their X chromosome are more likely to be affected because they have no second copy of the gene to compensate. Females, on the other hand, have two copies of the X chromosome, so they can be carriers of the mutated gene without showing symptoms. If a female is a carrier and has a child with a male who has the disorder, there is a 50% chance of passing on the mutated gene to their offspring, regardless of the child's sex.

2. **X-linked dominant inheritance:** In this pattern, a mutation in a gene on the X chromosome leads to the expression of a dominant trait or disorder. Both males and females can be affected by X-linked dominant disorders, but they may show different patterns of inheritance. Males with the mutated gene are often more severely affected because they have only one X chromosome. Females can be affected if they inherit the mutated gene from either parent. If a female with an X-linked dominant disorder has a child with an unaffected male, there is a 50% chance of passing on the mutated gene to each child, regardless of the child's sex.

It's important to note that while sex-linked inheritance is often associated with the X chromosome, the Y chromosome also carries genes that are passed down from fathers to their sons. These Y-linked genes are responsible for determining male-specific characteristics and traits.

Examples of X-linked disorders include hemophilia, Duchenne muscular dystrophy, and red-green color blindness. These disorders predominantly affect males, while females can be carriers without exhibiting symptoms.

Pedigree analysis:

Pedigree analysis in genetics involves the study of inherited traits and genetic disorders within families. It helps geneticists understand the patterns of inheritance and predict the likelihood of specific traits or diseases being passed onto future generations. Pedigree analysis is particularly useful in studying traits or disorders that have a genetic component.

Here are some key concepts and terms associated with pedigree analysis in genetics:

1. **Autosomal Inheritance:** Traits or disorders that are inherited on autosomal chromosomes (non-sex chromosomes) follow specific inheritance patterns. Autosomal dominant inheritance means that an affected individual has a 50% chance of passing the trait to each of their offspring. Autosomal recessive inheritance requires

both parents to be carriers or affected for the trait to be expressed in the offspring.

2. **X-Linked Inheritance:** Some traits or disorders are located on the X chromosome. X-linked dominant inheritance means that an affected male passes the trait to all his daughters, while an affected female can pass the trait to both sons and daughters. X-linked recessive inheritance means that the trait is more commonly expressed in males, as they have only one X chromosome.

3. **Pedigree Symbols:** Pedigree charts use standardized symbols to represent individuals and their relationships, as mentioned in the previous response. These symbols help track the presence or absence of a trait or disorder across generations.

4. **Consanguinity:** In some pedigrees, individuals who are closely related through a common ancestor may marry and have children. This is known as consanguinity. Consanguineous marriages increase the likelihood of inheriting rare recessive disorders because the parents share a larger proportion of their genetic material.

5. **Genetic Counseling:** Pedigree analysis plays a crucial role in genetic counseling. Genetic counselors use pedigree analysis to assess the risk of genetic disorders in families, provide information about inheritance patterns, and guide individuals or couples in making informed decisions about family planning.

By studying the patterns of inheritance within a family using pedigree analysis, geneticists can gain insights into the genetic basis of traits and disorders, develop hypotheses about the underlying genes involved, and contribute to the understanding and management of genetic conditions.

Genetic disorders (autosomal and sex-linked):

Genetic disorders are conditions that result from abnormalities or mutations in an individual's genes or chromosomes. These disorders can be classified into different types, including autosomal disorders and sex-linked disorders, based on their inheritance patterns.

1. **Autosomal Disorders:** Autosomal disorders are caused by mutations in genes located on the autosomes, which are non-sex chromosomes (chromosome pairs 1-22 in humans). Autosomal disorders can be further categorized into autosomal dominant disorders and autosomal recessive disorders.

- **Autosomal Dominant Disorders:** In autosomal dominant disorders, a single copy of the mutated gene is sufficient to cause the disorder. The affected gene can be inherited from an affected parent or may arise spontaneously. Examples of autosomal dominant disorders include Huntington's disease, Marfan syndrome, and neurofibromatosis.
- **Autosomal Recessive Disorders:** Autosomal recessive disorders require the inheritance of two copies of the mutated gene, one from each parent, to manifest the disorder. Carriers of the mutated gene typically do not show symptoms. Examples of autosomal recessive disorders include cystic fibrosis, sickle cell anemia, and Tay-Sachs disease.

2. **Sex-Linked Disorders:** Sex-linked disorders, also known as X-linked disorders, occur due to mutations in genes located on the sex chromosomes, particularly the X chromosome. Since males have one X and one Y chromosome, they are more susceptible to X-linked disorders because they lack a second X chromosome to compensate for the mutated gene.

- **X-Linked Dominant Disorders:** In X-linked dominant disorders, a single copy of the mutated gene on the X chromosome can cause the disorder. These disorders are more commonly observed in females since they have two X chromosomes. Examples include Rett syndrome and vitamin D-resistant rickets.
- **X-Linked Recessive Disorders:** X-linked recessive disorders require the inheritance of the mutated gene on the X

chromosome from both parents for the disorder to be expressed in males. Females can be carriers of the mutated gene without showing symptoms. Examples include hemophilia, Duchenne muscular dystrophy, and color blindness.

It's important to note that while autosomal disorders can affect both males and females equally, sex-linked disorders are more prevalent in males due to their unique sex chromosome composition. However, females can still be carriers and pass the mutated gene to their offspring. Genetic counseling and testing can help assess the risk of inheriting or passing on these disorders.

Genetic counseling and prenatal testing:

Genetic counseling and prenatal testing are important components of reproductive healthcare that help individuals and couples make informed decisions about their reproductive choices and potential genetic conditions in their offspring. Let's explore each of these topics in more detail:

1. **Genetic Counseling:** Genetic counseling involves the assessment and communication of information about genetic conditions and their implications. It is typically provided by healthcare professionals known as genetic counselors who are trained in medical genetics and counseling techniques.

During a genetic counseling session, the counselor will collect detailed medical and family history to assess the risk of genetic disorders or conditions. They may also order genetic tests if necessary. The counselor then interprets the results of these tests and provides information about the inheritance patterns, the likelihood of passing on a genetic condition, and available options for managing or treating the condition.

Genetic counseling can be helpful for individuals or couples who are planning to have children, have concerns about their family's genetic history, or have had previous pregnancies or children with genetic conditions. It can also assist individuals in understanding their own genetic risks for certain conditions and making informed decisions about healthcare and screening options.

2. **Prenatal Testing** : Prenatal testing involves a range of diagnostic and screening tests conducted during pregnancy to assess the health and development of the fetus. These tests aim to detect potential genetic or chromosomal abnormalities in the developing baby.

There are two main categories of prenatal testing:

a. **Screening Tests:** These tests provide information about the likelihood of a fetus having a particular genetic condition. They are non-invasive and carry little to no risk for the mother or the fetus. Common screening tests include blood tests (such as the first-trimester combined screening and the quad screen) and ultrasound examinations (such as nuchal translucency screening).

b. **Diagnostic Tests:** If a screening test indicates an increased risk of a genetic condition or if there are other concerns, diagnostic tests can be performed to confirm the diagnosis. These tests carry a small risk of complications, including miscarriage. Common diagnostic tests include chorionic villus sampling (CVS) and amniocentesis. These tests involve collecting a small sample of fetal cells or amniotic fluid to analyze the genetic material directly.

Prenatal testing allows parents to make informed decisions about the continuation of the pregnancy, prepare for the birth of a child with special needs, and, in some cases, explore treatment options or plan for medical interventions after birth.

It is important to note that decisions about genetic counseling and prenatal testing are personal and should be based on individual circumstances, values, and beliefs. Healthcare providers and genetic counselors play a vital role in guiding individuals and couples through the decision-making process and providing support and information throughout the journey.

Chapter 4:
Molecular Genetics

·   **DNA replication, transcription, and translation**

- Genetic code and protein synthesis
- Gene regulation and control
- Mutations and their consequences

Introduction:

Molecular genetics is a branch of genetics that focuses on studying the structure, function, and behavior of genes at the molecular level. It delves into the intricate workings of DNA, the molecule that carries genetic information, and explores how variations in genes can lead to different traits or diseases. By investigating the processes of DNA replication, gene expression, and regulation, molecular genetics provides insights into the fundamental mechanisms underlying inheritance and gene function.

Through the examination of DNA structure, molecular genetics reveals the double helix arrangement of nucleotide building blocks that form the genetic code. It unravels the mechanisms of DNA replication, ensuring accurate transmission of genetic information during cell division.

Gene expression, a key area of study in molecular genetics, involves the transcription of DNA into messenger RNA (mRNA), which is then translated into proteins. This field investigates the regulatory factors and processes involved in gene expression, shedding light on how cells interpret and utilize genetic information.

Molecular genetics also explores genetic mutations, which are alterations in the DNA sequence that can lead to changes in gene function. By understanding the different types of mutations and their consequences, researchers can gain insights into the causes of genetic disorders and devise potential treatments.

The study of gene regulation is another significant aspect of molecular genetics. It investigates the complex mechanisms through which genes are switched on or off in response to internal and external cues. This regulation is crucial for processes like cell differentiation, development, and response to environmental stimuli.

The field of molecular genetics has paved the way for genetic engineering, enabling scientists to manipulate and modify genes for various purposes. Techniques like recombinant DNA technology and gene editing tools such as CRISPR-Cas9 have revolutionized research in this field and opened doors to applications in medicine, agriculture, and biotechnology.

In summary, molecular genetics provides a detailed understanding of the molecular basis of inheritance, gene expression, and regulation. It has profound implications for our understanding of genetic diseases, development of therapies, and advancements in biotechnology. By unraveling the complexities of genes at the molecular level, molecular genetics contributes to our knowledge of life's fundamental processes.

DNA replication, transcription, and translation:

DNA replication, transcription, and translation are fundamental processes in molecular biology that are essential for the synthesis of proteins in living organisms. Let's explore each process in more detail:

1. **DNA Replication:** DNA replication is the process by which a cell duplicates its DNA to ensure that each daughter cell receives a complete set of genetic information. It occurs during the S (synthesis) phase of the cell cycle. The steps involved in DNA replication are as follows:

   a. **Initiation:** The replication begins at specific sites called origins of replication, where the DNA strands unwind and separate, forming a replication bubble.

   b. **Elongation:** Enzymes called DNA polymerases add complementary nucleotides to each of the separated DNA strands. The leading strand is synthesized continuously, while the lagging strand is synthesized in short fragments called Okazaki fragments.

   c. **Termination:** Once the entire DNA molecule has been replicated, specific termination signals halt the replication process, and the newly synthesized DNA molecules separate from each other.

2. **Transcription:** Transcription is the process by which genetic information encoded in DNA is used to synthesize an RNA molecule. It involves the transfer of genetic information from DNA to RNA. The steps involved in transcription are as

follows:

a. **Initiation:** RNA polymerase binds to a specific region of DNA called the promoter, marking the beginning of transcription.

b. **Elongation:** RNA polymerase synthesizes a complementary RNA molecule using one of the DNA strands as a template. The RNA molecule is synthesized in the 5' to 3' direction, and it is antiparallel to the DNA template strand.

c. **Termination:** Transcription continues until a termination signal is reached. At this point, RNA polymerase and the newly synthesized RNA molecule dissociate from the DNA template.

3. **Translation:** Translation is the process by which the information carried by RNA is used to synthesize proteins. It occurs in the cytoplasm and involves the participation of ribosomes and transfer RNA (tRNA) molecules. The steps involved in translation are as follows:

a. Initiation: The small ribosomal subunit binds to the mRNA molecule, and the initiator tRNA carrying the amino acid methionine binds to the start codon (usually AUG) on the mRNA. The large ribosomal subunit then joins the complex.

b. Elongation: The ribosome moves along the mRNA molecule, and tRNA molecules carrying specific amino acids bind to the codons on the mRNA, forming a polypeptide chain. Peptide bonds are formed between adjacent amino acids, and the ribosome moves from one codon to the next.

c. Termination: Translation continues until a stop codon (UAA, UAG, or UGA) is reached on the mRNA. At this point, the ribosome releases the completed polypeptide chain, and the components dissociate from each other.

These processes are essential for the accurate replication of DNA, the transcription of genetic information into RNA, and the subsequent translation of RNA into proteins, ultimately determining the structure and function of living organisms.

Genetic code and protein synthesis:

The genetic code and protein synthesis are fundamental processes in molecular biology that govern how genetic information is stored, transmitted, and utilized by living organisms. Let's start with the genetic code.

Genetic Code: The genetic code refers to the set of rules by which the information in DNA or RNA is translated into the amino acid sequence of a protein. It is essentially a mapping between a sequence of nucleotides (building blocks of DNA and RNA) and the corresponding amino acids (building blocks of proteins). The genetic code is universal, meaning it is nearly identical in all living organisms, from bacteria to humans.

The genetic code is composed of codons, which are sequences of three nucleotides. Each codon specifies a particular amino acid or a stop signal that marks the end of protein synthesis. There are 64 possible codons, and they encode for 20 standard amino acids, leaving some redundancy in the code. For example, the amino acid methionine is encoded by the codon "AUG," while tryptophan is encoded by the codon "UGG."

Protein Synthesis: Protein synthesis is the process by which the genetic information stored in DNA is used to build proteins. It occurs in two main steps: transcription and translation.

1. **Transcription:** Transcription takes place in the nucleus of eukaryotic cells or the cytoplasm of prokaryotic cells. During transcription, an enzyme called RNA polymerase binds to a specific region on the DNA molecule called the promoter. The DNA molecule is then unwound, and one of the DNA strands serves as a template for RNA synthesis.

RNA polymerase adds complementary nucleotides (adenine, uracil, cytosine, and guanine) to the growing RNA chain, following the rules of base pairing (A-U, C-G). The resulting RNA molecule, called messenger RNA (mRNA), is a copy of the DNA sequence and carries the genetic information from the nucleus to the cytoplasm.

2. **Translation:** Translation occurs in the cytoplasm on ribosomes, which are cellular structures composed of RNA and proteins. It involves the conversion of the mRNA sequence into an amino acid sequence to form a protein.

The mRNA binds to a ribosome, and the process begins with the recognition of the start codon (AUG) by a specific initiator tRNA molecule carrying the amino acid methionine. Then, other tRNA molecules, each carrying a specific amino acid, bind to the ribosome in a complementary manner to the codons on the mRNA.

As the ribosome moves along the mRNA, amino acids are joined together in a specific order dictated by the mRNA codons. The ribosome catalyzes the formation of peptide bonds between adjacent amino acids, resulting in the elongation of the growing polypeptide chain. This continues until a stop codon is reached, signaling the termination of protein synthesis.

After translation, the newly synthesized protein may undergo additional modifications, such as folding, post-translational modifications, and targeting specific cellular compartments, to become fully functional.

In summary, the genetic code provides the rules for translating the DNA or RNA sequence into the amino acid sequence of a protein. Protein synthesis involves transcription, where mRNA is synthesized from DNA, and translation, where the mRNA is used as a template to synthesize a protein on ribosomes. These processes are essential for the expression of genetic information and the functioning of living organisms.

Gene regulation and control:

Gene regulation and control refers to the processes by which genes are turned on or off, resulting in the regulation of gene expression. Gene expression is the process through which genetic information encoded in DNA is used to create functional molecules such as proteins or RNA molecules.

Gene regulation is essential for the proper functioning of cells and organisms. It allows cells to respond to changes in their environment, differentiate into specialized cell types, and maintain homeostasis. Gene regulation can occur at various levels, including transcriptional, post-transcriptional, translational, and post-translational regulation.

Transcriptional regulation is the most common and fundamental form of gene regulation. It involves the control of gene expression at the level of transcription, where the DNA sequence is copied into RNA by an enzyme called RNA polymerase. Transcriptional regulation is primarily achieved through the binding of regulatory proteins, known as transcription factors, to specific DNA sequences in the regulatory regions of genes. These regulatory regions include promoters, enhancers, and silencers, which are located near the gene or at a distance.

Transcription factors can activate or repress gene expression by facilitating or inhibiting the binding of RNA polymerase to the promoter region of a gene. They can also interact with other proteins to form complexes that modulate transcriptional activity. The activity of transcription factors is often influenced by signals from the cell's environment, such as hormones, nutrients, or developmental cues, allowing cells to adjust their gene expression profiles accordingly.

Post-transcriptional regulation occurs after the synthesis of RNA molecules. It involves processes such as alternative splicing, RNA editing, and the stability and degradation of RNA molecules. Alternative splicing allows the generation of multiple protein isoforms from a single gene by selectively including or excluding different exons during RNA processing. RNA stability and degradation are regulated by various factors that can determine the lifespan of RNA molecules.

Translational regulation influences the efficiency of translation, where RNA molecules are used as templates to synthesize proteins. It can be controlled through regulatory elements within the RNA molecule, such as the presence of specific sequences or structures that affect ribosome binding or initiation of translation.

Post-translational regulation involves modifications to proteins after they have been synthesized. These modifications can include phosphorylation, acetylation, glycosylation, or proteolytic cleavage, among others. Such modifications can alter the activity, localization, or stability of proteins, thereby regulating their function.

Overall, gene regulation and control are highly complex processes that involve a combination of genetic, biochemical, and cellular mechanisms. The interplay between these mechanisms allows cells and organisms to precisely control gene expression and adapt to their ever-changing internal and external environments.

Mutations and their consequences:

In genetics, mutations are changes or alterations that occur in the DNA sequence of an organism. These changes can happen due to various factors such as errors during DNA replication, exposure to mutagens (chemicals or radiation), or as a result of natural processes like recombination. Mutations can occur in different regions of the DNA, including genes, regulatory regions, or non-coding regions. They can have a range of consequences, which can be categorized as follows:

1. **Silent Mutations:** These mutations do not result in any significant change in the amino acid sequence of the protein. They occur when the altered codon still codes for the same amino acid, often due to the degeneracy of the genetic code.

2. **Missense Mutations:** These mutations result in the substitution of one amino acid for another in the protein sequence. Depending on the specific amino acid change and its location within the protein, missense mutations can have varying effects on protein structure and function. They can be benign, neutral, or deleterious.

3. **Nonsense Mutations:** These mutations lead to the formation of a premature stop codon in the DNA sequence. As a result, translation is terminated prematurely, leading to a truncated protein that is usually non-functional.

4. **Frameshift Mutations:** These mutations involve the addition or deletion of nucleotides in the DNA sequence, causing a shift in the reading frame during translation. Frameshift mutations typically result in a completely altered amino acid sequence downstream of the mutation site, often leading to a non-functional protein.

5. **Insertions and Deletions (Indels):** These mutations involve the insertion or deletion of one or more nucleotides in the DNA sequence. Depending on the size and location of the

indel, they can have significant effects on protein structure and function.

6. **Chromosomal Rearrangements:** Mutations can also involve larger-scale alterations in the structure of chromosomes. Examples include inversions, translocations, duplications, and deletions of chromosomal segments. These rearrangements can disrupt gene function, alter gene expression patterns, or lead to the creation of fusion genes with novel properties.

The consequences of mutations can vary widely, from no apparent effect to severe phenotypic consequences. Some mutations can be beneficial, providing organisms with advantages in specific environments or contributing to evolutionary processes. Others can be detrimental, leading to genetic disorders, increased susceptibility to diseases, or developmental abnormalities. The impact of a mutation depends on factors such as the type of mutation, its location within the genome, the affected gene's function, and the genetic background of the organism.

Chapter 5:
Population Genetics

· **Hardy-Weinberg equilibrium**
· **Factors affecting gene frequency (selection, genetic drift, gene flow)**
· **Genetic variation and human populations**
· **Speciation and evolutionary processes**

Introduction:

Population genetics is a field of study within biology that focuses on understanding how genetic variation and evolutionary processes shape and change the genetic makeup of populations over time. It combines principles from genetics, evolution, and statistics to investigate patterns and mechanisms of genetic diversity within and between populations.

At its core, population genetics examines the genetic variation within a group of individuals, known as a population, and investigates how this variation arises and is maintained. It explores the factors that influence the distribution and frequency of genetic variants, such as mutations, genetic drift, gene flow, natural selection, and recombination.

One of the central concepts in population genetics is the Hardy-Weinberg equilibrium, which describes the relationship between the frequencies of different genetic variants and the processes that can disrupt this equilibrium. Deviations from the Hardy-Weinberg equilibrium can indicate the action of evolutionary forces and provide insights into the dynamics of population genetic processes. By studying patterns of genetic variation, population geneticists can address a wide range of questions related to evolutionary biology, conservation genetics, human genetics, and even forensics. They may investigate how populations adapt to changing environments, examine the genetic basis of diseases, trace human migration patterns, or analyze the impact of human activities on wildlife populations.

Population genetics has important implications for understanding the genetic basis of diversity and the mechanisms driving evolutionary change. By unraveling the processes that shape genetic variation, population genetics contributes to our understanding of the biological world and has practical applications in fields such as medicine, agriculture, and conservation.

Hardy-Weinberg equilibrium:

Hardy-Weinberg equilibrium is a fundamental concept in population genetics that describes the theoretical genetic structure of a population in the absence of evolutionary forces. It provides a baseline against which changes in genetic frequencies can be measured. The equilibrium was formulated by G.H. Hardy and W. Weinberg independently in 1908.

The Hardy-Weinberg equilibrium is based on several assumptions:

1. **Large population size:** The population must be large enough for genetic drift to have a minimal effect on allele

frequencies. Random fluctuations in smaller populations can lead to significant changes in allele frequencies over time.

2. **Random mating:** Individuals in the population must mate randomly, with no preference for particular genotypes. Non-random mating, such as assortative mating (mating based on phenotype similarity), can disrupt the equilibrium.

3. **No mutation:** The alleles in the population remain constant and do not change due to new mutations. Mutations introduce new alleles into a population, which can alter the equilibrium.

4. **No migration:** There is no migration into or out of the population. Migration can introduce new alleles or remove existing ones, leading to changes in genetic frequencies.

5. **No natural selection:** There are no selective pressures acting on the population. In the absence of natural selection, all genotypes have equal fitness and contribute equally to the next generation.

Under these assumptions, the Hardy-Weinberg equilibrium predicts that the genotype frequencies of a population will remain constant from generation to generation. It defines the relationship between allele frequencies (p and q) and genotype frequencies (p^2, 2pq, and q^2), where p represents the frequency of one allele and q represents the frequency of the other allele in a diploid population. The equilibrium can be expressed by the Hardy-Weinberg equation:
$p^2 + 2pq + q^2 = 1$
Where:

- p^2 represents the frequency of individuals homozygous for allele p.
- 2pq represents the frequency of individuals heterozygous for alleles p and q.
- q^2 represents the frequency of individuals homozygous for allele q.

By comparing observed genotype frequencies in a population to the expected frequencies calculated from the Hardy-Weinberg equation, researchers can identify deviations and infer the presence of evolutionary forces such as natural selection, migration, genetic drift, or non-random mating.

Factors affecting gene frequency (selection, genetic drift, gene flow):

In genetics, several factors can influence the frequency of genes within a population. Three important factors are natural selection, genetic drift, and gene flow. These forces can shape the genetic composition of a population over time.

1. **Natural selection:** Natural selection is the process by which certain genetic traits become more or less common in a population due to their influence on an organism's survival and reproductive success. If a particular genetic variation provides a fitness advantage, individuals with that variation are more likely to survive and reproduce, passing on the beneficial genes to future generations. Over time, this can lead to an increase in the frequency of the advantageous genes in the population.

2. **Genetic drift:** Genetic drift refers to random changes in gene frequencies that occur in small populations due to chance events. In small populations, random sampling errors can have a significant impact on gene frequencies from one generation to the next. Genetic drift tends to be more pronounced in small populations and can lead to the loss or fixation of certain genes. Unlike natural selection, genetic drift is a random process and does not depend on the fitness of the genes.

3. **Gene flow:** Gene flow refers to the movement of genes between different populations. It occurs when individuals migrate and breed with individuals from another population, introducing new genetic material. Gene flow can increase genetic diversity within a population and reduce genetic differences between populations. It can counteract the effects of genetic drift and natural selection by introducing new variations and preventing populations from becoming genetically isolated.

These factors are not mutually exclusive and often interact with each other. Natural selection can act on genetic variations that arise from mutation or are introduced through gene flow. Genetic drift can have a more significant impact in small populations or during population bottlenecks. Gene flow can counterbalance the effects of genetic drift by introducing new genetic material and maintaining genetic diversity. Understanding these factors is crucial for studying how populations evolve and adapt to their environments.

Genetic variation and human populations:

Genetic variation refers to the diversity of genetic material within a population or species. It is the result of differences in the DNA sequences of individuals, which can manifest as variations in traits, susceptibility to diseases, and other genetic characteristics.

Human populations exhibit genetic variation due to several factors. These include:

1. **Mutation:** Mutations are random changes in the DNA sequence that can occur during DNA replication or as a result of environmental factors. Mutations can introduce new genetic variants into a population.

2. **Genetic recombination:** During sexual reproduction, genetic material from two parents is combined, resulting in offspring with a unique combination of genetic variants. This process shuffles and recombines genetic information, increasing genetic diversity within a population.

3. **Migration:** Human populations have migrated and interacted with each other throughout history, leading to the exchange of genetic material. This process, known as gene flow, introduces new genetic variants into populations and can reduce genetic differences between populations.

4. **Genetic drift:** Genetic drift refers to random changes in the frequency of genetic variants within a population over time. It is more pronounced in small populations and can lead to the loss or fixation of specific genetic variants, reducing genetic diversity.

Human populations have historically been geographically isolated, leading to the development of distinct genetic characteristics in different regions. Over time, however, populations have mixed through migration and interbreeding, reducing the genetic differences between populations. Nonetheless, some genetic variation still exists among human populations due to factors such as regional adaptations, historical population movements, and natural selection acting on specific traits.

Understanding genetic variation and its distribution across human populations is essential for studying human evolution, genetic diseases, personalized medicine, and forensic genetics. However, it is important to approach the topic with caution to avoid misinterpretation or the promotion of harmful ideas related to race or genetic superiority, as genetic variation does not define human worth or intelligence.

Speciation and evolutionary processes:

Speciation is the process by which new species arise from a common ancestral population. It is an important concept in evolutionary biology and genetics. Speciation occurs when populations of the same species become reproductively isolated from each other, preventing gene flow between them.

There are several mechanisms that can lead to speciation, and they can be broadly categorized into two main types: allopatric speciation and sympatric speciation.

1. **Allopatric Speciation:** This type of speciation occurs when populations become geographically isolated from each other. Geographic barriers such as mountains, rivers, or islands can physically separate populations, preventing gene flow. Over time, genetic differences can accumulate due to mutations, genetic drift, and natural selection in each isolated population. Eventually, if the genetic differences become significant enough, the populations may no longer be able to interbreed even if the physical barriers are removed.

2. **Sympatric Speciation:** Sympatric speciation occurs when new species arise within the same geographic area without any physical barriers to gene flow. This process is often driven by non-geographic factors such as ecological

specialization or sexual selection. For example, if a population becomes adapted to a specific ecological niche or evolves a unique mating behavior or preference, it may become reproductively isolated from the rest of the population, leading to speciation.

In both allopatric and sympatric speciation, genetic changes play a crucial role in driving the divergence of populations. Mutations, which are random changes in DNA, provide the raw material for genetic variation. Genetic drift, the random fluctuation of allele frequencies in small populations, can also have a significant impact on genetic divergence, especially in small or isolated populations. Natural selection acts on these genetic variations, favoring traits that enhance survival and reproduction in specific environments, leading to adaptation and potentially the formation of new species. Additionally, other processes such as gene flow, genetic recombination, and hybridization can also influence speciation. Gene flow occurs when individuals from different populations mate and exchange genes, potentially counteracting the genetic differences that lead to speciation. Genetic recombination, which occurs during sexual reproduction, can shuffle and combine genetic material from different individuals, creating new genetic combinations. Hybridization, the interbreeding between two different species, can lead to the formation of hybrid individuals that may have different genetic traits and reproductive barriers from their parent species.

Overall, speciation and evolutionary processes in genetics involve a complex interplay of genetic variation, mutation, genetic drift, natural selection, gene flow, genetic recombination, and other factors. These processes shape the genetic diversity of populations and ultimately contribute to the formation of new species over time.

Chapter 6:

Chromosomal Basis of Inheritance

- Chromosome structure and organization
- Sex determination and sex chromosomes
- Chromosomal aberrations (aneuploidy, polyploidy)
- Linkage and recombination

Introduction:

The chromosomal basis of inheritance is a fundamental concept in the field of genetics that explains how genes are passed from parents to offspring. It provides an understanding of how traits are inherited and how variations occur in populations.

At the core of this concept is the understanding that genes, which carry the hereditary information, are located on chromosomes. Chromosomes are thread-like structures found in the nucleus of cells and are composed of DNA and proteins. Humans typically have 23 pairs of chromosomes, with one set inherited from each parent. The connection between chromosomes and inheritance was first established by the pioneering work of Gregor Mendel in the 19th century, but the specific role of chromosomes in inheritance became clearer in the early 20th century through the research of Thomas Hunt Morgan and his colleagues.

Morgan and his team conducted experiments using the fruit fly Drosophila melanogaster and observed that certain traits, such as eye color, were inherited together. They discovered that these traits were linked to specific chromosomes. Furthermore, they noticed that some traits did not follow the expected patterns of inheritance, which led to the discovery of sex-linked traits.

The sex chromosomes, designated as X and Y, play a significant role in the chromosomal basis of inheritance. In humans, females have two X chromosomes (XX), while males have one X and one Y chromosome (XY). Genes located on the X and Y chromosomes exhibit distinct patterns of inheritance due to their differences in structure and number.

An important concept related to the chromosomal basis of inheritance is genetic recombination. During meiosis, the process of cell division that produces gametes (sperm and egg cells), homologous chromosomes pair up and exchange genetic material through a process called crossing over. This leads to the shuffling and recombination of genes, resulting in genetic diversity among offspring.

Additionally, chromosomal abnormalities can occur due to errors during meiosis or chromosomal rearrangements. These abnormalities, such as aneuploidy (abnormal number of chromosomes) or structural changes, can have significant effects on an individual's health and development.

In summary, the chromosomal basis of inheritance explains how genes are organized on chromosomes and transmitted from generation to generation. It also elucidates the role of chromosomes in determining the inheritance of traits, including the impact of sex chromosomes and the occurrence of genetic recombination. Understanding this concept is crucial for comprehending the mechanisms of genetic inheritance and the variability observed in populations.

Chromosome structure and organization:

Chromosomes are structures within cells that contain the genetic material of an organism. They are made up of DNA (deoxyribonucleic acid) molecules tightly packaged with proteins called histones. The structure and organization of chromosomes play a crucial role in the transmission of genetic information from one generation to the next.

Here's a breakdown of chromosome structure and organization:

1. **DNA:** The main component of chromosomes is DNA, which carries the genetic instructions for the development, functioning, and reproduction of all living organisms. DNA is a long, double-stranded molecule made up of four nucleotide bases: adenine (A), thymine (T), cytosine (C), and

guanine (G). The sequence of these bases encodes the genetic information.

2. **Histones:** DNA molecules wrap around proteins called histones, forming a complex called chromatin. Histones play a crucial role in packaging and organizing DNA. They act as spools around which the DNA is wound, allowing for efficient storage of genetic material.

3. **Nucleosomes:** The basic unit of chromatin is the nucleosome, which consists of a segment of DNA wrapped around a core of eight histone proteins. The nucleosomes are connected by linker DNA, resulting in a "beads on a string" structure.

4. **Chromatid:** During the cell division process, chromosomes become condensed and visible under a microscope. Each chromosome consists of two identical copies called chromatids, which are held together at a region called the centromere. The chromatids contain the exact same DNA sequence and are formed during DNA replication.

5. **Homologous Chromosomes:** In diploid organisms (such as humans), chromosomes are organized in pairs called homologous chromosomes. One chromosome in each pair is inherited from the mother, and the other is inherited from the father. Homologous chromosomes have the same genes at corresponding positions, although they may have different alleles (variants) of those genes.

6. **Karyotype:** The complete set of chromosomes in an organism is called its karyotype. In humans, the karyotype consists of 46 chromosomes, with 22 pairs of autosomes (non-sex chromosomes) and one pair of sex chromosomes (XX in females and XY in males). The specific arrangement and organization of chromosomes in a karyotype are unique to each species.

Overall, the structure and organization of chromosomes allow for the efficient packaging, replication, and transmission of genetic information. Changes or abnormalities in chromosome structure can lead to genetic disorders or developmental abnormalities. The study of chromosomes and their organization is essential in the field of genetics and plays a crucial role in understanding inheritance patterns and genetic diseases.

Sex determination and sex chromosomes:

Sex determination in genetics refers to the process by which an organism's biological sex is determined. In most species, including mammals, sex is determined by the presence of specific sex chromosomes. These chromosomes carry the genetic information that determines whether an individual will develop as a male or a female.

In humans, as well as in many other mammals, sex determination is based on the presence or absence of the Y chromosome. The sex chromosomes in humans are labeled as X and Y. Females typically have two X chromosomes (XX), while males have one X and one Y chromosome (XY). The Y chromosome contains genes that initiate the development of male characteristics.

During reproduction, each parent contributes one sex chromosome to their offspring. The mother always contributes an X chromosome, while the father can contribute either an X or a Y chromosome. If the father contributes an X chromosome, the resulting offspring will be female (XX). If the father contributes a Y chromosome, the offspring will be male (XY).

The presence of the Y chromosome triggers the development of male reproductive structures and characteristics. It contains a gene called the SRY (sex-determining region Y) gene, which is responsible for initiating male development. This gene stimulates the production of testosterone, a hormone that directs the formation of male genitalia and other male secondary sexual characteristics during embryonic development.

It's important to note that while sex determination is typically associated with the presence of specific chromosomes, there are exceptions in different species. For example, some reptiles have temperature-dependent sex determination, where the incubation temperature of the eggs determines the sex of the offspring. Other organisms, such as birds, have a different set of sex chromosomes (ZW for females and ZZ for males).

In summary, sex determination in genetics is the process by which an individual's biological sex is determined, usually based on the presence or absence of specific sex chromosomes. In humans, the presence of a Y chromosome triggers male development, while the absence of a Y chromosome leads to female development.

Chromosomal aberrations (aneuploidy, polyploidy:

Chromosomal aberrations refer to structural or numerical abnormalities in chromosomes. Chromosomes are the structures within cells that carry genetic information in the form of DNA. Normally, humans have 46 chromosomes arranged in 23 pairs, with 22 pairs of autosomes (non-sex chromosomes) and one pair of sex chromosomes (XX in females and XY in males). Chromosomal aberrations can result in alterations to this normal chromosome structure or number.

Aneuploidy is a type of chromosomal aberration characterized by an abnormal number of chromosomes in a cell. It occurs when there is a gain or loss of one or more chromosomes, resulting in an imbalance of genetic material. For example, trisomy is a type of aneuploidy where there is an extra copy of a chromosome, such as trisomy 21 (Down syndrome), where there are three copies of chromosome 21 instead of the usual two. Monosomy is another type of aneuploidy, which involves the loss of one copy of a chromosome. A well-known example is Turner syndrome, where females have only one X chromosome instead of the usual two (45,X).

Polyploidy, on the other hand, is a chromosomal aberration characterized by an increase in the number of complete sets of chromosomes. It occurs when there is a duplication of the entire chromosome set. Polyploidy is more commonly observed in plants than in animals. Triploidy (3n), where there are three sets of chromosomes, and tetraploidy (4n), where there are four sets of chromosomes, are examples of polyploidy. Polyploid organisms often exhibit different characteristics compared to their diploid counterparts and can have altered fertility and reproductive behaviors.

Chromosomal aberrations, including aneuploidy and polyploidy, can have significant effects on an organism's development and function. They can cause various genetic disorders, birth defects, and reproductive issues. The severity of these effects depends on the specific chromosomes involved and the extent of the imbalance in genetic material. Chromosomal aberrations can arise due to errors during DNA replication, chromosome segregation, or exposure to certain environmental factors or chemicals.

Linkage and recombination:

In genetics, linkage and recombination are two important concepts that describe the inheritance patterns of genes located on the same chromosome. Let's break down each concept separately:

1. **Linkage** : Linkage refers to the tendency of genes located close together on the same chromosome to be inherited together as a unit. When genes are tightly linked, they are less likely to undergo independent assortment during meiosis, the process of cell division that produces reproductive cells (gametes) such as sperm and eggs. As a result, the linked genes tend to be inherited as a package, maintaining their association across generations.

The degree of linkage between genes is measured by the recombination frequency, which is the percentage of recombinant offspring produced in a cross. Recombinant offspring carry new combinations of alleles that differ from those of their parents due to recombination events during meiosis.

2. **Recombination:** Recombination, also known as genetic recombination or crossing over, is the exchange of genetic material between homologous chromosomes during meiosis. It occurs in a region called the chiasma, where nonsister chromatids of homologous chromosomes break and rejoin.

Recombination plays a crucial role in generating genetic diversity within a population. It shuffles the alleles (alternative forms of genes) on homologous chromosomes, creating new combinations of alleles that were not present in the parental chromosomes. This process is responsible for the independent assortment of genes and allows for the creation of unique genetic combinations in offspring. The frequency of recombination between two genes is influenced by the physical distance between them on the chromosome. Genes that are located farther apart are more likely to undergo recombination, whereas genes that are closely located (tightly linked) have a lower chance of recombination.

By studying linkage and recombination patterns, geneticists can create genetic maps that depict the relative positions of genes on chromosomes. These maps provide valuable information for understanding inheritance patterns, gene mapping, and identifying genetic diseases.

Chapter 7:
Biotechnology and Genetic Engineering

- Recombinant DNA technology
- DNA cloning and genetic transformation
- Genetically modified organisms (GMOs)
- Applications in medicine, agriculture, and industry

Introduction:

Biotechnology and genetic engineering are closely related fields that involve the manipulation of living organisms and their genetic material to develop new products, improve processes, and solve problems in various sectors, including medicine, agriculture, and industry.

Biotechnology is a broad term that refers to the use of living organisms or their components to create or modify products, improve plants or animals, or develop new technologies. It encompasses a wide range of techniques and applications, including genetic engineering.

Genetic engineering, also known as genetic modification or recombinant DNA technology, specifically focuses on altering an organism's genetic material by introducing or modifying specific genes. This process involves manipulating DNA, the molecule that carries the genetic instructions for the development, functioning, and reproduction of living organisms.

The key steps involved in genetic engineering are as follows:

1. **Identification of a desired trait:** Scientists identify a specific trait or characteristic they want to introduce or modify in an organism. This could include traits such as disease resistance in crops, increased production of a specific protein, or the ability to produce a useful chemical.

2. **Isolation of genes:** The genes responsible for the desired trait are identified and isolated from the DNA of another organism. These genes can come from the same species (intraspecies transfer) or a different species (interspecies transfer).

3. **Gene insertion:** The isolated genes are inserted into the DNA of the target organism. This is typically done using specialized techniques such as gene guns, microinjection, or bacterial vectors. The goal is to integrate the new genes into the target organism's genome so that they are replicated and passed on to subsequent generations.

4. **Selection and screening:** The genetically modified organisms (GMOs) are screened and selected based on the successful incorporation of the desired genes. Various techniques like DNA analysis, polymerase chain reaction (PCR), and other molecular biology methods are used to identify and confirm the presence of the introduced genes.

5. **Expression of the modified genes:** Once the genes are successfully integrated, they are expressed by the target organism. This means that the genetic instructions carried by the inserted genes are used to produce specific proteins or enzymes that confer the desired trait.

The applications of biotechnology and genetic engineering are diverse and include:

1. **Medicine:** Genetic engineering has revolutionized the production of pharmaceuticals, including insulin, growth hormones, and vaccines. It is also used in gene therapy to treat genetic disorders by replacing or correcting faulty genes.
2. **Agriculture:** Genetic engineering is employed to develop genetically modified crops with improved characteristics, such as higher yield, pest resistance, drought tolerance, and enhanced nutritional value. It can also be used to create crops with specific traits, such as the ability to produce edible vaccines.
3. **Industrial applications:** Biotechnology is utilized in the production of enzymes, biofuels, biodegradable plastics, and various other industrial chemicals. Microorganisms are often genetically modified to enhance their ability to produce these substances efficiently.
4. **Environmental applications:** Bioremediation, which involves using living organisms to clean up pollutants and contaminants in the environment, is an important application of biotechnology. Genetic engineering can be used to modify organisms to make them more effective in breaking down or removing harmful substances.

It is worth noting that the field of biotechnology and genetic engineering is subject to extensive regulation and ethical considerations due to the potential risks and impacts associated with the release of genetically modified organisms into the environment and the use of genetic manipulation in humans.

Recombinant DNA technology:

Recombinant DNA technology, also known as genetic engineering or gene splicing, is a technique used in genetics to manipulate and modify the DNA (deoxyribonucleic acid) of an organism. It involves combining genetic material from different sources, typically from different species, to create a recombinant DNA molecule. This technique allows scientists to transfer specific genes or genetic traits between organisms that would not naturally breed or exchange genetic material.

The process of recombinant DNA technology involves several steps:

1. **Isolation of DNA:** The DNA containing the gene of interest is extracted from the donor organism. This can be done using various methods, such as cell lysis and purification techniques.
2. **Cutting DNA:** The DNA molecule is then cut at specific points using restriction enzymes. These enzymes act as molecular scissors and recognize specific DNA sequences, cutting the DNA at those sites. This generates fragments of DNA with sticky ends or cohesive ends.
3. **Cloning vector preparation:** A cloning vector, typically a plasmid or a viral vector, is prepared to receive the foreign DNA. The vector is also cut with the same restriction enzymes, generating complementary sticky ends.
4. **DNA ligation:** The donor DNA fragments are then combined with the cut cloning vector, and DNA ligase is used to join the fragments. The sticky ends of the donor DNA and the vector base pair, creating recombinant DNA molecules.
5. **Introduction into host organism:** The recombinant DNA molecule is introduced into a host organism, such as bacteria, yeast, or plants. This is done using various techniques, including transformation, transfection, or injection.
6. **Selection and cloning:** The host organism is then selected based on specific markers present in the vector, such as antibiotic resistance genes. This allows only the organisms containing the desired recombinant DNA to survive and reproduce. The host organism is cloned, producing multiple copies of the recombinant DNA.
7. **Expression of the gene:** If the goal is to produce a specific protein encoded by the inserted gene, the host organism is allowed to express the gene. The recombinant DNA is transcribed and translated into a functional protein.

Recombinant DNA technology has numerous applications in genetics and biotechnology. It enables the production of valuable proteins, such as insulin and growth hormones, through the expression of genes in host organisms. It also allows for the modification of crops to exhibit desirable traits, such as disease resistance or increased yield. Recombinant DNA technology has revolutionized the field of genetics and has had a profound impact on medicine, agriculture, and other scientific endeavors.

DNA cloning and genetic transformation:

DNA cloning and genetic transformation are two fundamental techniques used in molecular biology and genetic engineering to manipulate and study genes.

1. **DNA Cloning** : DNA cloning is the process of making identical copies of a specific DNA fragment or gene. It involves inserting the DNA fragment of interest into a vector, which is typically a small, self-replicating piece of DNA, such as a plasmid. The vector serves as a carrier to propagate the DNA fragment in a host organism, usually a bacterium or yeast.

The process of DNA cloning typically involves the following steps:

a. **Isolation of the DNA fragment:** The DNA fragment of interest is obtained from a source organism using various extraction techniques.

b. **Selection of a vector:** A suitable vector, such as a plasmid, is chosen. Vectors often contain genes that confer antibiotic resistance, which allows for the selection of host cells that have successfully taken up the vector.

c. **Insertion of DNA into the vector:** The DNA fragment and the vector are cut with specific enzymes (restriction enzymes) that generate compatible ends. These ends are then joined together using an enzyme called DNA ligase, resulting in a recombinant DNA molecule.

d. **Introduction of the recombinant DNA into host cells:** The recombinant DNA molecule is introduced into host cells using techniques such as transformation, electroporation, or viral transduction.

e. **Selection of transformed cells:** Host cells that have successfully taken up the recombinant DNA are identified by selecting for the antibiotic resistance gene present in the vector.

f. **Cloning and propagation:** The transformed cells are grown in a culture medium, allowing them to multiply and form a colony of genetically identical cells, each containing a copy of the cloned DNA fragment.

2. **Genetic Transformation:** Genetic transformation refers to the process of introducing foreign DNA into an organism's genome, thereby changing its genetic makeup. This technique allows researchers to introduce new traits or modify existing ones in an organism.

The process of genetic transformation usually involves the following steps:

a. **Selection of the target organism:** The organism to be transformed is chosen based on its amenability to genetic manipulation and the desired application.

b. **Isolation and modification of the gene of interest:** The gene of interest, which encodes the desired trait, is isolated and modified as required. Modifications can involve altering the gene sequence or introducing additional regulatory elements to control gene expression.

c. **Delivery of foreign DNA into the host organism:** Various methods can be used to introduce the modified DNA into the target organism's cells, such as direct injection, microinjection, electroporation, or the use of viral vectors.

d. **Integration of foreign DNA into the host genome:** Once inside the host cells, the foreign DNA integrates into the genome through recombination or other mechanisms, becoming a part of the host organism's genetic material.

e. **Expression of the introduced gene:** The integrated gene is transcribed and translated by the host organism's cellular machinery, resulting in the production of the desired protein or trait.

Genetic transformation allows researchers to study gene function, produce genetically modified organisms (GMOs) with desirable traits, develop disease models, and create biotechnological products such as pharmaceuticals, enzymes, or biofuels.

Both DNA cloning and genetic transformation are powerful tools in molecular biology and genetic engineering, enabling scientists to investigate gene function, create recombinant DNA molecules, and modify organisms for various purposes.

Genetically modified organisms (GMOs):

Genetically modified organisms (GMOs) are living organisms whose genetic material has been altered using modern biotechnology techniques, particularly genetic engineering. Genetic engineering involves the manipulation and transfer of specific genes or DNA segments from one organism to another, or even from one species to another, in order to introduce desired traits or characteristics.

The process of creating GMOs begins with the identification of a specific trait or characteristic found in one organism that is desired in another. For example, scientists may identify a gene in a plant that allows it to resist pests or tolerate certain environmental conditions, and they may want to transfer that gene into another plant species to confer the same traits.

Once the desired gene or DNA segment is identified, scientists use various techniques to isolate and extract the gene from the donor organism. This gene is then inserted into the genome of the recipient organism, often through the use of a vector, such as a plasmid or a virus. The vector helps deliver the gene into the recipient organism's cells, where it integrates into the genome and becomes a part of its genetic material.

The modified organism, now containing the inserted gene, is then grown or bred in a laboratory or controlled environment to allow it to express the desired traits. This could involve growing genetically modified crops, raising genetically modified animals, or cultivating genetically modified microorganisms.

GMOs have been developed for various purposes, including improving crop yield, enhancing nutritional content, increasing resistance to pests, diseases, or environmental stressors, and reducing the need for certain pesticides or herbicides. They have been used in agriculture, medicine, and industry.

The use of GMOs has sparked considerable debate and controversy. Critics express concerns about potential risks to human health and the environment, such as allergenicity, the development of pesticide resistance in pests, and the unintended effects on non-target organisms. Proponents argue that GMOs can help address global challenges in food security, agricultural sustainability, and public health.

It is worth noting that regulations regarding GMOs vary across countries, with some nations imposing strict labeling and safety assessment requirements, while others have more relaxed regulations. The development and use of GMOs continue to be an active area of scientific research, regulation, and public discourse.

Applications in medicine, agriculture, and industry:

Genetics plays a crucial role in various fields, including medicine, agriculture, and industry. Here's an overview of how genetics is applied in these domains:

1. **Medicine:**
   - **Disease Diagnosis:** Genetic testing allows the identification of genetic variations associated with diseases. This helps diagnose genetic disorders, predict disease risk, and enable personalized medicine.
   - **Pharmacogenomics:** By studying an individual's genetic makeup, pharmacogenomics helps determine how a person may respond to specific medications. This information aids in tailoring drug treatments to maximize efficacy and minimize side effects.
   - **Gene Therapy:** Genetic engineering techniques are used to introduce functional genes into cells to treat genetic disorders. It holds promise for curing diseases caused by faulty genes.
   - **Cancer Genetics:** Genetic analysis helps identify genetic mutations and variations associated with cancer. This information is valuable for cancer diagnosis, prognosis, and the development of targeted therapies.
2. **Agriculture:**

- **Crop Improvement:** Genetic engineering techniques are employed to develop genetically modified (GM) crops with desirable traits such as increased yield, pest resistance, drought tolerance, or enhanced nutritional value.
- **Livestock Improvement:** Genetic selection and breeding techniques are used to improve livestock by selecting animals with desired traits, such as increased milk production, disease resistance, or improved meat quality.
- **Disease Resistance:** Genetic markers are identified to select and breed plants and animals with resistance to diseases, reducing the need for pesticides or antibiotics.

3. **Industry:**
   - **Industrial Enzymes:** Genetic engineering enables the production of enzymes used in various industries, such as detergent production, biofuel production, and food processing. Genetically modified microorganisms can be engineered to produce large quantities of specific enzymes efficiently.
   - **Bioremediation:** Genetically engineered organisms can be designed to degrade pollutants and toxins, aiding in the cleanup of contaminated environments.
   - **Industrial Fermentation:** Genetic modifications can enhance microorganisms used in fermentation processes, leading to improved yields, faster production, and better-quality products in industries like brewing, winemaking, and bioethanol production.

These are just a few examples of how genetics is applied in medicine, agriculture, and industry. Advances in genetic research and technology continue to expand the possibilities for improving human health, increasing food production, and enhancing industrial processes.

Chapter 8:
Genomics and Personalized Medicine

- **Human genome project and sequencing technologies**
- **Comparative genomics and functional genomics**
- **Pharmacogenomics and precision medicine**
- **Ethical considerations in genomics research**

Introduction:

Genomics and personalized medicine are two fields that have revolutionized the field of genetics and healthcare. Genomics refers to the study of an organism's entire set of genes, known as its genome. It involves analyzing the structure, function, and interactions of genes to understand their role in health and disease. Personalized medicine, on the other hand, is an approach to healthcare that tailors medical treatment to an individual's specific genetic makeup, lifestyle, and environment. It recognizes that each person is unique and responds differently to various treatments. By considering an individual's genetic information, personalized medicine aims to provide targeted and more effective treatments, as well as preventive strategies.

Advancements in genomic technologies, such as DNA sequencing, have significantly enhanced our understanding of the human genome. The Human Genome Project, completed in 2003, was a landmark scientific achievement that mapped the entire human genome. Since then, the cost of sequencing has dramatically decreased, making it more accessible and allowing for large-scale genomic studies.

The knowledge gained from genomics has led to important discoveries about the genetic basis of various diseases, including cancer, cardiovascular disorders, and rare genetic conditions. Genomic research has identified numerous genes and genetic variants associated with disease susceptibility, drug response, and treatment outcomes.

In personalized medicine, genomic information plays a crucial role in guiding treatment decisions. Genetic testing can identify specific genetic mutations or variations that influence an individual's response to certain medications. This information helps healthcare professionals determine the most appropriate drug and dosage for a particular patient, minimizing adverse reactions and optimizing therapeutic outcomes.

Moreover, genomics and personalized medicine are not limited to the treatment of diseases but also extend to disease prevention and risk assessment. By analyzing an individual's genomic profile, it is possible to identify individuals who are at a higher risk of developing certain conditions. This knowledge allows for targeted screening, early intervention, and lifestyle modifications to reduce the risk or delay the onset of diseases.

However, the implementation of genomics and personalized medicine faces various challenges. Privacy concerns, ethical considerations, and the need for robust data analysis and interpretation are among the key issues that need to be addressed. Additionally, ensuring equitable access to genomic technologies and incorporating genetic information into clinical practice remain ongoing areas of research and development.

Overall, genomics and personalized medicine have the potential to transform healthcare by providing tailored approaches to disease prevention, diagnosis, and treatment. As our understanding of the human genome continues to grow, these fields will likely play an increasingly significant role in shaping the future of medicine.

Human genome project and sequencing technologies:

The Human Genome Project (HGP) was an international research effort conducted between 1990 and 2003 with the goal of determining the complete sequence of the human genome. The genome refers to the entire set of genetic information present in an organism. The HGP successfully mapped and sequenced the approximately 3 billion base pairs that make up the human genome.

The completion of the HGP has had a profound impact on the field of genetics and has opened up numerous opportunities for further research and understanding of human biology. One of the major outcomes of the project was the identification of all human genes, which has provided valuable insights into the genetic basis of human traits, diseases, and susceptibility to certain conditions. Advancements in sequencing technologies played a crucial role in the success of the Human Genome Project and continue to drive progress in the field of genetics. There have been significant improvements in DNA sequencing methods since the completion of the project, leading to increased speed, accuracy, and reduced costs of sequencing.

Traditional Sanger sequencing, used during the Human Genome Project, has been largely replaced by next-generation sequencing (NGS) technologies. NGS methods allow for the parallel sequencing of millions of DNA fragments, enabling rapid and cost-effective sequencing of entire genomes. These technologies include platforms such as Illumina's HiSeq and NovaSeq systems, Ion Torrent's Personal Genome Machine (PGM), and Pacific Biosciences' Single Molecule, Real-Time (SMRT) sequencing.

Another important development in sequencing technology is the emergence of single-molecule sequencing methods, which can generate long reads of DNA sequences. Technologies such as Oxford Nanopore's MinION and PacBio's Sequel systems offer the ability to sequence long stretches of DNA without the need for fragmentation, providing valuable information for genome assembly and structural variant analysis.

Sequencing technologies have not only facilitated the study of human genetics but have also been applied to a wide range of research areas, including cancer genomics, microbiology, evolutionary biology, and personalized medicine. The ability to rapidly sequence genomes has enabled the identification of disease-causing mutations, the discovery of new genes and genetic variants, and the exploration of complex genetic interactions.

Overall, the Human Genome Project and subsequent advancements in sequencing technologies have revolutionized the field of genetics, allowing researchers to gain a deeper understanding of the human genome and its implications for health and disease. These advancements continue to drive new discoveries and innovations in the field, shaping our understanding of genetics and paving the way for personalized medicine and targeted therapies.

Comparative genomics and functional genomics:

Comparative genomics and functional genomics are two important fields in genetics that contribute to our understanding of the structure, function, and evolution of genomes. While they are distinct disciplines, they are often intertwined and complement each other in studying the relationship between genotype and phenotype.

1. **Comparative Genomics:** Comparative genomics involves comparing the genomes of different species to identify similarities, differences, and patterns of genomic evolution. It provides insights into the evolutionary relationships between organisms and helps to uncover the genetic basis of various traits and functions.

Key aspects of comparative genomics include:

a. **Genome Assembly and Annotation:** Comparative genomics relies on the availability of high-quality genome sequences for multiple organisms. These genomes are assembled and annotated to identify genes, regulatory regions, and other functional elements.

b. **Orthology and Homology:** Comparative genomics uses the concepts of orthology and homology to identify genes that are derived from a common ancestral gene. Orthologous genes are present in different species and share a common ancestry, whereas paralogous genes arise from gene duplication events within a species.

c. **Evolutionary Analysis:** By comparing genomes, scientists can study the evolutionary changes that have occurred over time. This includes identifying conserved regions, gene gain and loss events, and evolutionary innovations that have shaped the genomes of different organisms.

d. **Phylogenomics:** Comparative genomics also aids in constructing phylogenetic trees, which depict the evolutionary relationships between organisms. By comparing genetic sequences, such as genes or whole genomes, researchers can infer the relatedness of species and their divergence points.

2. **Functional Genomics:** Functional genomics focuses on understanding the functions of genes and other functional elements within a genome. It aims to decipher how genetic information is utilized to carry out biological processes and contribute to the phenotype of an organism.

Key aspects of functional genomics include:

a. **Gene Expression Profiling:** Functional genomics investigates gene expression patterns by analyzing the transcriptome (all the RNA molecules transcribed from genes) using techniques like microarrays or RNA sequencing. This helps identify which genes are active in specific tissues, developmental stages, or under different conditions.

b. **Gene Function Annotation:** Functional genomics aims to assign functions to genes and other genomic elements. This involves experimental techniques like knockout or knockdown studies to observe the effects of gene inactivation or reduction in expression. Additionally, computational approaches can predict gene functions based on sequence similarity, protein domain analysis, and network analysis.

c. **Regulatory Element Analysis:** Functional genomics explores the role of regulatory elements, such as promoters, enhancers, and non-coding RNAs, in controlling gene expression. Techniques like chromatin immunoprecipitation sequencing (ChIP-seq) and DNA footprinting help identify and characterize these elements.

d. **Functional Screens:** Functional genomics employs large-scale screening approaches, such as RNA interference (RNAi) or CRISPR-Cas9-based knockout screens, to systematically investigate the function of genes. These screens can identify genes involved in specific biological processes or disease pathways.

Overall, comparative genomics provides a broad evolutionary context and helps identify conserved elements, while functional genomics focuses on unraveling the specific functions and mechanisms of genes and other genomic elements. By combining the two approaches, researchers gain a comprehensive understanding of the relationships between genome structure, function, and evolution.

Pharmacogenomics and precision medicine:

Pharmacogenomics and precision medicine are two related fields that utilize genetic information to personalize medical treatment. Pharmacogenomics refers to the study of how an individual's genetic makeup influences their response to drugs. It combines the disciplines of pharmacology (the study of how drugs work in the body) and genomics (the study of an organism's complete set of genes). The goal of pharmacogenomics is to identify genetic variations that affect drug metabolism, efficacy, and adverse reactions. By understanding these genetic factors, healthcare providers can tailor drug treatments to individual patients, maximizing effectiveness and minimizing the risk of adverse effects. Precision medicine, on the other hand, is a broader concept that takes into account various factors, including genetics, lifestyle, environment, and medical history, to deliver personalized healthcare. It aims to provide the right treatment to the right patient at the right time. In the context of genetics, precision medicine utilizes genetic information to guide treatment decisions and optimize outcomes. It recognizes that individuals differ in their response to medications due to genetic variations and seeks to identify these variations to inform treatment choices.

The integration of pharmacogenomics into precision medicine allows healthcare providers to select drugs and doses that are most likely to be effective and safe for a particular patient based on their genetic profile. By considering genetic factors, physicians can avoid drugs that are likely to be ineffective or cause adverse reactions in certain individuals. This approach helps to improve patient outcomes, reduce adverse drug reactions, and optimize medication use.

Pharmacogenomic testing involves analyzing a patient's DNA to identify specific genetic variants that may influence drug response.

This information is then used to guide treatment decisions. For example, a genetic test may reveal that an individual metabolizes a certain medication more slowly than average due to a genetic variant in a specific enzyme. Armed with this knowledge, a healthcare provider can adjust the dosage to ensure the drug is effective without causing toxicity.

Overall, pharmacogenomics and precision medicine in genetics aim to enhance the effectiveness and safety of medical treatments by considering individual genetic variations. By tailoring treatments to a person's genetic profile, healthcare providers can optimize therapeutic outcomes and improve patient care.

Ethical considerations in genomics research:

Ethical considerations play a crucial role in genomics research, especially in the field of genetics study. Here are some key ethical considerations that researchers need to address:

1. **Informed Consent:** Researchers must obtain informed consent from individuals participating in genomics research. This means that participants should be fully aware of the purpose, risks, benefits, and potential implications of the study before giving their consent. Informed consent ensures that participants have autonomy and can make decisions about their involvement based on accurate information.

2. **Privacy and Confidentiality:** Genomics research involves the collection and analysis of individuals' genetic information, which is highly personal and sensitive. Researchers must ensure that participant data is protected and that privacy and confidentiality are maintained. Adequate measures should be in place to safeguard data against unauthorized access or disclosure, and data should be de-identified whenever possible to minimize the risk of re-identification.

3. **Data Sharing and Access:** While protecting participant privacy, there is also a need to balance data sharing for the advancement of scientific knowledge. Researchers should consider the benefits of sharing genomic data with other scientists and ensure that appropriate mechanisms are in

place for responsible data sharing. This may involve establishing data access agreements, setting guidelines for data use, and ensuring that data sharing does not lead to potential harms or discrimination against individuals.

4. **Equity and Justice:** Genomics research should be conducted in a manner that promotes fairness, justice, and equity. Researchers need to be mindful of potential biases or discrimination that may arise from the analysis and interpretation of genetic information. Special attention should be given to ensure that research findings are not misused to perpetuate societal inequalities, stigmatization, or discrimination based on genetic characteristics.

5. **Benefit and Risk Assessment:** Researchers should carefully evaluate the potential benefits and risks associated with genomics research. The potential benefits may include advancements in personalized medicine, disease prevention, and improved understanding of human biology. However, researchers must also identify and minimize potential risks, such as psychological distress, discrimination, and unintended consequences resulting from the use of genetic information.

6. **Community Engagement and Collaboration:** Engaging with the communities affected by genomics research is essential. Researchers should involve diverse stakeholders, including patient groups, communities, and ethicists, in the design, implementation, and interpretation of genomics studies. Collaborative approaches help ensure that research is culturally sensitive, respects community values, and addresses specific concerns or priorities.

7. **Regulatory Compliance:** Researchers must adhere to applicable laws, regulations, and ethical guidelines related to genomics research. Institutional review boards (IRBs) or ethics committees play a vital role in reviewing research protocols and ensuring compliance with ethical standards. Researchers should follow established guidelines for

responsible conduct, data management, and reporting of results.

These ethical considerations aim to protect the rights and welfare of research participants, ensure the responsible and equitable conduct of genomics research, and foster public trust in scientific advancements in the field of genetics.

Chapter 9:
Gene Expression and Regulation

· **Transcription factors and gene regulation networks**
· **Epigenetics and chromatin remodeling**
· **Non-coding RNAs (microRNAs, long non-coding RNAs)**
· **Gene expression profiling techniques**

Introduction:

Gene expression refers to the process by which genetic information encoded in DNA is utilized to produce functional proteins or RNA molecules within a cell. It involves a series of complex molecular events that allow genes to be "expressed" or activated, resulting in the synthesis of specific proteins that perform various cellular functions.

Gene regulation, on the other hand, refers to the mechanisms by which gene expression is controlled. It allows cells to finely tune the levels of gene expression in response to different internal and external cues, ensuring proper development, growth, and functioning of organisms. Gene regulation occurs at multiple levels, including transcriptional, post-transcriptional, translational, and post-translational levels.

Transcriptional regulation is the primary mode of gene regulation, which occurs during the process of transcription. Transcription is the synthesis of an RNA molecule from a DNA template, catalyzed by an enzyme called RNA polymerase. Transcriptional regulation involves the binding of various regulatory proteins, called transcription factors, to specific DNA sequences near the genes of interest. These transcription factors can either enhance (activators) or inhibit (repressors) the transcription of genes, thereby controlling their expression.

Post-transcriptional regulation involves the processing and modification of RNA molecules after transcription. It includes mechanisms such as alternative splicing, where different exons of a gene are selectively included or excluded from the final RNA transcript, leading to the generation of multiple protein isoforms from a single gene. Other post-transcriptional mechanisms include RNA editing, RNA stability control, and the action of small regulatory RNA molecules like microRNAs.

Translational regulation refers to the control of gene expression at the level of protein synthesis. It involves the regulation of ribosome binding to messenger RNA (mRNA) molecules, which determines the rate at which proteins are produced from the mRNA transcripts. Various factors, such as RNA structure, translation initiation factors, and microRNAs, can influence translational regulation.

Post-translational regulation involves the modification and regulation of proteins after their synthesis. This includes processes like protein folding, protein modifications (e.g., phosphorylation, acetylation), and protein degradation, which collectively influence the stability, activity, and localization of proteins within the cell.

Studying gene expression and regulation is crucial for understanding the molecular mechanisms underlying normal biological processes and diseases. Techniques such as DNA microarrays, RNA sequencing (RNA-seq), and chromatin immunoprecipitation (ChIP) enable researchers to measure gene expression levels, identify regulatory elements, and study the interactions between regulatory proteins and DNA. These studies provide valuable insights into the complex networks of gene regulation and can help in the development of novel therapies and interventions targeting specific genes or regulatory pathways.

Transcription factors and gene regulation networks:

Transcription factors (TFs) and gene regulation networks play crucial roles in genetic studies, particularly in understanding how genes are regulated and how they contribute to various biological processes. Let's break down these concepts:

1. **Transcription Factors (TFs):** Transcription factors are proteins that bind to specific DNA sequences called transcription factor binding sites (TFBS) located in the promoter or enhancer regions of genes. TFs regulate gene expression by controlling the rate of transcription, which is the process of copying DNA into RNA. By binding to the TFBS, TFs can either activate (enhance) or repress (inhibit) the transcription of target genes.

TFs contain specific DNA-binding domains that allow them to recognize and bind to specific DNA sequences. Once bound, they can recruit other proteins and complexes to the gene's regulatory regions, such as co-activators or co-repressors, which further modulate gene expression. Through this mechanism, TFs act as molecular switches that control when and to what extent genes are turned on or off.

2. **Gene Regulation Networks:** Gene regulation networks refer to the intricate interconnections between transcription factors, genes, and other regulatory elements in a cell or organism. These networks describe the complex interactions that determine gene expression patterns and govern various cellular processes.

In a gene regulation network, TFs can regulate multiple genes, and a single gene can be regulated by multiple TFs. These interactions create a network of regulatory relationships, which can be represented as graphical models or mathematical equations to understand and predict gene expression behavior.

Advances in high-throughput technologies such as next-generation sequencing and microarrays have enabled researchers to generate large-scale data sets to analyze gene expression profiles and identify TFs involved in specific biological processes or diseases. By integrating this information with computational and statistical methods, scientists can construct gene regulation networks that provide insights into the regulatory mechanisms underlying complex biological systems.

Studying gene regulation networks helps us understand how genes work together, how their expression is coordinated, and how disruptions in these networks can lead to diseases or developmental abnormalities. It also aids in identifying potential therapeutic targets and designing interventions to modulate gene expression for medical applications.

Overall, transcription factors and gene regulation networks are fundamental concepts in genetic studies that provide valuable insights into the complexity of gene expression regulation and its impact on cellular processes and disease states.

Epigenetics and chromatin remodeling:

Epigenetics and chromatin remodeling are two interconnected areas of study within the field of genetics that provide valuable insights into gene regulation and the inheritance of traits. Let's explore each of these concepts individually.

1. **Epigenetics:** Epigenetics refers to the study of heritable changes in gene expression that occur without any alterations to the underlying DNA sequence. It involves modifications to the structure of DNA or the associated proteins that regulate gene activity. These modifications can be influenced by environmental factors and can have long-lasting effects on gene expression.

The primary mechanisms of epigenetic regulation include DNA methylation and histone modification. DNA methylation involves the addition of a methyl group to the DNA molecule, typically at cytosine residues. Methylation at certain gene regions, such as gene promoters, often leads to gene silencing or reduced gene expression.

Histone modification, on the other hand, refers to chemical alterations of the proteins called histones, which form the structural backbone of chromatin. Histone modifications include acetylation, methylation, phosphorylation, and others, which can either promote or inhibit gene expression by altering the accessibility of DNA to the transcriptional machinery.

Epigenetic modifications can be passed on from one generation to another, leading to the inheritance of specific gene expression

67

patterns. However, they can also be reversible, allowing for dynamic responses to environmental stimuli.

2. **Chromatin Remodeling:** Chromatin remodeling refers to the dynamic alteration of the structure and composition of chromatin, which is the complex of DNA and proteins that make up the chromosomes. Chromatin remodeling complexes facilitate changes in the positioning of nucleosomes (histone-DNA complexes) along the DNA, altering the accessibility of specific gene regions.

These complexes can slide nucleosomes along the DNA, evict nucleosomes from specific sites, or replace histones with variant histone proteins. By doing so, they regulate the accessibility of the DNA to transcription factors and other regulatory proteins, thereby influencing gene expression.

Chromatin remodeling plays a critical role in various cellular processes, including gene activation, gene repression, DNA repair, and genome stability. It ensures that the right genes are expressed at the right time and in the right cells.

Studying epigenetics and chromatin remodeling helps researchers understand how genes are regulated and how specific traits are inherited. It provides insights into the mechanisms underlying various diseases, including cancer and developmental disorders. Moreover, it offers potential avenues for therapeutic interventions by targeting epigenetic modifications and chromatin remodeling processes to modify gene expression patterns.

Non-coding RNAs (microRNAs, long non-coding RNAs):

Non-coding RNAs (ncRNAs) are a class of RNA molecules that do not code for proteins but play crucial roles in various biological processes. Two important types of ncRNAs extensively studied in genetic research are microRNAs (miRNAs) and long non-coding RNAs (lncRNAs).

1. **MicroRNAs (miRNAs):** MicroRNAs are short, single-stranded RNA molecules typically consisting of around 22 nucleotides. They are transcribed from specific genes in the genome but do not encode proteins. Instead, they regulate gene expression at the post-transcriptional level by binding

to messenger RNA (mRNA) molecules. The binding of miRNAs to their target mRNAs leads to mRNA degradation or inhibition of translation, thereby preventing the production of the corresponding protein.

MicroRNAs have been found to have diverse functions and are involved in numerous biological processes, including development, cell differentiation, immune responses, and disease pathogenesis. Dysregulation of miRNAs has been implicated in various disorders, including cancer, cardiovascular diseases, neurodegenerative diseases, and metabolic disorders. Therefore, studying miRNAs provides insights into gene regulation networks and can help identify potential therapeutic targets.

2. **Long non-coding RNAs (lncRNAs):** Long non-coding RNAs are RNA molecules that are longer than 200 nucleotides and do not encode proteins. They are transcribed from specific genomic regions and can be found in various subcellular compartments, including the nucleus, cytoplasm, and organelles. LncRNAs exhibit complex and diverse mechanisms of action and can interact with DNA, RNA, and proteins.

LncRNAs have been implicated in a wide range of biological processes, including chromatin remodeling, gene transcription, RNA processing, and protein localization. They play critical roles in cellular development, differentiation, and disease progression. Dysregulated expression of lncRNAs has been associated with various diseases, including cancer, cardiovascular disorders, and neurological conditions. Additionally, lncRNAs can serve as biomarkers for disease diagnosis and prognosis.

The study of miRNAs and lncRNAs in genetics involves various techniques such as microarray analysis, RNA sequencing (RNA-seq), and functional assays. These approaches help identify differentially expressed ncRNAs, investigate their targets, and explore their biological functions. Additionally, computational methods and bioinformatics tools are employed to predict miRNA target genes and lncRNA-protein interactions, facilitating the understanding of their regulatory mechanisms.

In summary, microRNAs and long non-coding RNAs are essential players in genetic studies. They contribute to gene regulation networks, impact various biological processes, and have significant implications in disease development and progression. Understanding their functions and mechanisms provides valuable insights into the complex landscape of gene expression and offers potential avenues for therapeutic interventions.

Gene expression profiling techniques:

Gene expression profiling is a set of techniques used in genetic studies to measure the activity of genes within a cell or tissue. It provides valuable insights into how genes are regulated and how they contribute to various biological processes and diseases. There are several methods available for gene expression profiling, each with its own strengths and limitations. Here are some commonly used techniques:

1. **Microarrays:** Microarrays allow researchers to simultaneously measure the expression levels of thousands of genes. The technique involves immobilizing short DNA or RNA probes representing specific genes onto a solid surface, such as a glass slide or a microchip. The labeled RNA or cDNA from the sample is then hybridized to the probes, and the signal intensity is measured to determine the expression levels of the corresponding genes.

2. **RNA-Seq:** RNA-Seq (RNA sequencing) is a high-throughput technique that provides a comprehensive and quantitative analysis of gene expression. It involves converting RNA molecules from the sample into complementary DNA (cDNA) and then sequencing the cDNA fragments using next-generation sequencing technologies. The resulting sequence reads are aligned to a reference genome or assembled de novo to identify and quantify the expressed genes.

3. **Quantitative PCR (qPCR):** qPCR is a widely used technique for the accurate quantification of gene expression. It involves the amplification of specific RNA or DNA targets using fluorescently labeled primers and a DNA polymerase enzyme. The fluorescence signal is measured during each

amplification cycle, and the threshold cycle (Ct) is determined, which correlates with the initial amount of the target transcript in the sample.

4. **NanoString technology:** NanoString is a hybridization-based method that allows for the direct detection and quantification of specific RNA molecules without amplification. It uses color-coded molecular barcodes to capture and count individual RNA molecules. The captured RNA targets are then detected and quantified using a digital analyzer.

5. **Single-cell RNA sequencing:** Single-cell RNA sequencing (scRNA-seq) provides gene expression profiles at the level of individual cells, allowing for the characterization of cellular heterogeneity within a tissue or organism. This technique involves isolating and sequencing RNA from individual cells, which provides insights into cell types, cell states, and gene regulatory networks.

These are just a few examples of gene expression profiling techniques used in genetic studies. Each technique has its own advantages and limitations, and the choice of method depends on the specific research question, available resources, and the desired level of resolution and sensitivity. Gene expression profiling techniques have revolutionized our understanding of gene regulation and have enabled the discovery of novel biomarkers and therapeutic targets in various diseases.

Chapter 10:
Genetic Techniques and Tools

· **Polymerase chain reaction (PCR)**
· **DNA sequencing methods**
· **Gel electrophoresis and DNA fingerprinting**
· **Bioinformatics and computational biology**

Introduction:

Genetic techniques and tools refer to a set of methodologies and technologies used in the field of genetics to study and manipulate genes, DNA, and genetic information. These techniques have revolutionized our understanding of genetics and have practical applications in various fields, including medicine, agriculture, and biotechnology. Here's a brief introduction to some commonly used genetic techniques and tools:

1. **Polymerase Chain Reaction (PCR):** PCR is a widely used technique that amplifies a specific DNA sequence, allowing researchers to generate millions of copies of a particular DNA fragment. It is essential for various applications, such as DNA sequencing, gene expression analysis, and genetic fingerprinting.

2. **DNA Sequencing:** DNA sequencing allows scientists to determine the precise order of nucleotides (A, T, C, G) in a DNA molecule. Next-generation sequencing (NGS) technologies have revolutionized this field, enabling rapid and cost-effective sequencing of entire genomes.

3. **Gene Cloning:** Gene cloning involves the isolation and replication of a specific DNA sequence, resulting in multiple copies of the same gene. This technique allows scientists to study the structure and function of genes and produce large quantities of gene products, such as proteins.

4. **Genetic Engineering:** Genetic engineering involves the manipulation of an organism's genetic material to introduce new traits or modify existing ones. This technique typically involves the insertion of foreign DNA into an organism's genome, enabling the production of desired proteins or the alteration of specific characteristics.

5. **CRISPR-Cas9:** CRISPR-Cas9 is a revolutionary gene editing tool that allows scientists to precisely modify genes within living organisms. It utilizes a molecular complex that can be programmed to target and edit specific DNA sequences, offering unprecedented control over genetic modifications.

6. **RNA Interference (RNAi) :** RNAi is a technique used to silence or inhibit the expression of specific genes. By

introducing small RNA molecules that target and degrade the corresponding messenger RNA (mRNA), RNAi can selectively "turn off" genes and study their functions.

7. **Genomic Editing:** Genomic editing techniques, such as CRISPR-based systems, have emerged as powerful tools for modifying the DNA sequences of organisms. These techniques enable precise editing of genomes, offering possibilities for correcting genetic diseases, enhancing crop traits, and advancing basic research.

These are just a few examples of the wide range of genetic techniques and tools available to researchers. They have significantly advanced our understanding of genetics and have the potential to revolutionize fields such as medicine, agriculture, and biotechnology.

Polymerase chain reaction (PCR):

Polymerase chain reaction (PCR) is a powerful molecular biology technique used in genetic studies to amplify specific segments of DNA. It was invented by Kary Mullis in 1983 and has since revolutionized various fields of research, including genetics, forensics, and medical diagnostics.

PCR allows scientists to selectively amplify and generate multiple copies of a target DNA sequence in a test tube. This process is achieved through a series of temperature-dependent steps that are repeated multiple times, resulting in exponential amplification of the desired DNA.

The basic steps involved in a PCR reaction are as follows:

1. **Denaturation:** The double-stranded DNA template is heated to a high temperature (typically around 95°C), causing the hydrogen bonds between the DNA strands to break, resulting in the separation of the double-stranded DNA into two single strands.

2. **Annealing:** The temperature is then lowered (typically around 50-65°C), allowing short DNA primers to bind specifically to complementary sequences on each of the separated single DNA strands. These primers flank the region of interest that needs to be amplified.

3. **Extension/Elongation:** The temperature is raised to an optimal range for DNA polymerase activity (typically around 72°C). A heat-stable DNA polymerase enzyme, such as Taq polymerase, adds complementary nucleotides to the primers, synthesizing new DNA strands that are complementary to the single-stranded DNA template.

These three steps (denaturation, annealing, and extension) make up a single cycle of PCR. Each cycle doubles the amount of the target DNA, resulting in an exponential increase in the number of DNA copies.

PCR is typically performed in a thermal cycler, a laboratory instrument that can rapidly and precisely control the temperature changes required for each step. The number of cycles can vary depending on the starting amount of DNA and the desired final amount of amplified product. Generally, 20 to 40 cycles are sufficient to generate a significant quantity of the target DNA.

PCR is a highly versatile technique used in various genetic studies. It allows researchers to amplify specific regions of DNA, such as genes or regulatory sequences, from a complex mixture of DNA. Amplified DNA can be further analyzed through techniques like DNA sequencing, restriction enzyme digestion, or genetic profiling methods.

Overall, PCR has become an indispensable tool in genetic research, enabling the detection and analysis of specific DNA sequences with high sensitivity and specificity.

DNA sequencing methods:

DNA sequencing is a crucial technique in genetic studies that allows researchers to determine the precise order of nucleotides (adenine, thymine, cytosine, and guanine) in a DNA molecule. The information obtained from DNA sequencing is fundamental for understanding various biological processes, identifying genetic variations, and studying the genetic basis of diseases.

There are several methods of DNA sequencing that have been developed over the years, with each method having its own advantages, limitations, and applications. Here are a few commonly used DNA sequencing methods:

1. **Sanger sequencing (also known as chain termination sequencing):** This was the first method developed for DNA sequencing. It involves the use of dideoxynucleotides (ddNTPs), which lack the 3' hydroxyl group necessary for DNA chain elongation. In Sanger sequencing, a DNA template is divided into four separate reactions, each containing a small amount of a specific ddNTP. The ddNTPs are incorporated randomly during DNA synthesis, resulting in the termination of DNA elongation at various positions. The resulting DNA fragments are separated by size using gel electrophoresis, and the sequence is determined by analyzing the pattern of terminated fragments.

2. **Next-generation sequencing (NGS):** NGS refers to a group of high-throughput sequencing technologies that revolutionized DNA sequencing by enabling the parallel sequencing of multiple DNA fragments. NGS methods include technologies such as Illumina sequencing, Ion Torrent sequencing, and Pacific Biosciences sequencing. These methods involve fragmenting the DNA into smaller pieces, attaching adapters or primers to the fragments, and amplifying them to create millions or billions of copies. The amplified fragments are then sequenced simultaneously using various techniques, and the resulting sequence reads are aligned and assembled to reconstruct the original DNA sequence.

3. **Third-generation sequencing:** Third-generation sequencing methods, such as single-molecule real-time (SMRT) sequencing developed by Pacific Biosciences and nanopore sequencing developed by Oxford Nanopore Technologies, offer long-read sequencing capabilities. These methods can sequence much longer DNA fragments, providing valuable information about structural variations, repetitive sequences, and complex genomic regions. SMRT sequencing employs a DNA polymerase that incorporates fluorescently labeled nucleotides while monitoring changes in light intensity, whereas nanopore sequencing passes DNA strands

through nanopores and measures changes in electrical current as the DNA moves through the pore.

4. **Single-cell sequencing:** Traditional sequencing methods typically require a substantial amount of DNA, which makes studying individual cells challenging. Single-cell sequencing techniques, such as single-cell RNA sequencing (scRNA-seq) and single-cell DNA sequencing (scDNA-seq), have emerged to address this limitation. These methods enable the analysis of gene expression or genomic variations within individual cells, providing insights into cellular heterogeneity and cell-to-cell differences in gene regulation or genetic alterations.

These are just a few examples of the DNA sequencing methods used in genetic studies. Each method has its own strengths and limitations in terms of read length, throughput, cost, and error rates. Researchers choose the appropriate sequencing method based on the specific research goals and the characteristics of the samples being studied.

Gel electrophoresis and DNA fingerprinting:

Gel electrophoresis and DNA fingerprinting are widely used techniques in genetic studies for analyzing and comparing DNA samples. Let's explore each of these methods in detail:

1. **Gel Electrophoresis:** Gel electrophoresis is a technique that separates DNA fragments based on their size and charge. The process involves creating an electric field across a gel matrix, usually made of agarose or polyacrylamide. DNA samples are loaded into wells at one end of the gel, and when an electric current is applied, the DNA molecules move through the gel.

The gel acts as a molecular sieve, with smaller DNA fragments moving faster and farther than larger ones. This separation occurs because DNA is negatively charged, and as the current passes through the gel, it attracts the DNA molecules towards the positive electrode (anode). The speed at which the DNA fragments move depends on their size and the porosity of the gel.

After the electrophoresis process, the DNA fragments appear as distinct bands on the gel. These bands can be visualized using various staining methods, such as ethidium bromide or fluorescent dyes. By comparing the positions of DNA bands in different samples, scientists can analyze the genetic information and study various aspects, such as genetic mutations, gene expression, or DNA profiling.

2. **DNA Fingerprinting:** DNA fingerprinting, also known as DNA profiling or genetic fingerprinting, is a technique used to identify individuals based on their unique genetic information. It utilizes highly variable regions of the genome, known as short tandem repeats (STRs), which are short sequences of DNA that repeat multiple times.

The process of DNA fingerprinting involves several steps:

a. **DNA Extraction:** DNA is extracted from the biological material, such as blood, saliva, or hair follicles, obtained from the individual or the crime scene.

b. **PCR Amplification:** Polymerase Chain Reaction (PCR) is used to selectively amplify specific STR regions in the DNA sample. Primers are designed to target these regions, and multiple cycles of DNA replication are performed to produce a sufficient amount of DNA for analysis.

c. **Gel Electrophoresis:** The amplified DNA fragments are then subjected to gel electrophoresis. The gel separates the DNA fragments based on their sizes, resulting in distinct bands for each STR region.

d. **DNA Visualization:** The gel is stained, and the DNA bands are visualized. The unique pattern of band sizes represents an individual's DNA fingerprint.

e. **Comparison and Analysis:** The DNA fingerprint of an individual is compared to other individuals or a DNA database to determine relationships, identify suspects, or establish genetic profiles.

DNA fingerprinting is commonly used in forensic science to match suspects to crime scene evidence. It is also used in paternity testing, immigration cases, and studying genetic relationships between individuals.

Overall, gel electrophoresis is a crucial technique for separating DNA fragments, while DNA fingerprinting uses this method to analyze and compare DNA samples for identification and genetic analysis purposes.

Bioinformatics and computational biology:

Bioinformatics and computational biology are interdisciplinary fields that merge biology, genetics, and computer science to analyze and interpret biological data, particularly in the context of genetic studies. They play a crucial role in understanding the complex information encoded within an organism's genetic material, such as DNA and RNA.

Genetic studies involve investigating the structure, function, and variations in genes and genomes to gain insights into various biological phenomena, including disease mechanisms, evolutionary relationships, and genetic traits. These studies generate vast amounts of data, such as DNA sequences, gene expression profiles, and protein structures, which require advanced computational techniques for analysis.

Bioinformatics is primarily concerned with developing and applying computational methods, algorithms, and tools to manage, organize, and analyze biological data. It involves the use of statistical and computational approaches to derive meaningful insights from large-scale genetic datasets. Bioinformaticians utilize various computational techniques, including sequence alignment, database searching, data mining, and machine learning, to extract knowledge from genetic information.

Computational biology, on the other hand, focuses on developing mathematical models and simulations to understand biological processes and phenomena. It combines computational and statistical approaches with experimental data to build models that describe the behavior of biological systems. Computational biologists employ techniques such as network analysis, mathematical modeling, and simulation to explore complex biological systems and predict their behavior.

In genetic studies, bioinformatics and computational biology play vital roles in several areas:

1. **Genome assembly and annotation:** Bioinformatics tools are used to piece together and annotate DNA sequences obtained from genome sequencing projects. These tools help identify genes, regulatory elements, and other functional regions within the genome.
2. **Sequence analysis:** Bioinformatics techniques enable the comparison of DNA and protein sequences to identify similarities, infer evolutionary relationships, and predict functional domains and motifs. Sequence alignment algorithms, such as BLAST, are widely used in this context.
3. **Gene expression analysis:** Computational methods are employed to analyze gene expression data obtained from techniques like microarrays or RNA sequencing. This analysis helps identify differentially expressed genes, discover regulatory networks, and understand gene function in specific biological processes.
4. **Structural biology:** Computational approaches are utilized to predict protein structures based on their amino acid sequences. This aids in understanding protein function, drug design, and protein-protein interactions.
5. **Systems biology:** Bioinformatics and computational biology contribute to the field of systems biology by integrating and analyzing large-scale biological data from multiple sources. This enables the construction of comprehensive models that capture the interactions and dynamics of biological systems.

Overall, bioinformatics and computational biology provide powerful tools and techniques for processing and interpreting genetic data, facilitating advancements in our understanding of genetics, disease mechanisms, and evolutionary processes. They play a crucial role in accelerating genetic research and have applications in fields such as personalized medicine, agriculture, and biotechnology.

Chapter 11:

Applied Genetics

- Medical genetics and genetic disorders
- Forensic genetics and DNA profiling
- Agricultural biotechnology and crop improvement
- Conservation genetics and biodiversity preservation

Introduction:

Applied genetics is a branch of genetics that focuses on the practical application of genetic knowledge and techniques in various fields, such as medicine, agriculture, and biotechnology. It involves using genetic information to understand and manipulate the characteristics of organisms for specific purposes.

In genetic studies, applied genetics plays a crucial role in advancing our understanding of inherited traits and genetic diseases. It involves the application of genetic principles and techniques to identify, analyze, and modify specific genes or genetic variations associated with particular traits or diseases.

In the field of medicine, applied genetics is used for genetic counseling, prenatal testing, and diagnosing and treating genetic disorders. Geneticists utilize techniques like DNA sequencing, gene expression analysis, and genome editing to identify disease-causing mutations and develop targeted therapies.

In agriculture, applied genetics is employed to enhance crop yields, improve resistance to pests and diseases, and develop genetically modified organisms (GMOs). It involves breeding programs to select and propagate desirable traits in plants and animals, including increased productivity, improved nutrition, and environmental adaptability.

Applied genetics also finds applications in forensic science, where DNA profiling techniques are used for criminal investigations and identification of individuals. It assists in establishing biological relationships, solving crimes, and exonerating innocent individuals.

Moreover, applied genetics contributes to advancements in biotechnology and pharmaceutical research. It enables the production of therapeutic proteins, the development of gene therapies, and the creation of genetically engineered organisms for various purposes.

Overall, applied genetics is a multidisciplinary field that integrates genetic knowledge with other scientific disciplines to address real-world challenges and improve our understanding of genetic traits, diseases, and their applications in diverse areas.

Medical genetics and genetic disorders:

Medical genetics is a branch of medicine that focuses on the study of genes, heredity, and genetic variation in relation to human health and disease. It involves the diagnosis, management, and counseling of individuals and families with genetic disorders or a predisposition to such disorders.

Genes are the basic units of heredity, containing the instructions for building and maintaining an organism. They are made up of DNA and are located on chromosomes within the nucleus of our cells. Each person inherits two copies of most genes, one from each parent.

Genetic disorders, also known as genetic diseases or inherited disorders, are conditions caused by changes or mutations in genes. These mutations can range from single-gene disorders, where a mutation in a single gene leads to a specific disease, to complex disorders influenced by multiple genes and environmental factors. There are thousands of known genetic disorders, each with its own set of symptoms, inheritance patterns, and underlying genetic mutations. Some examples of genetic disorders include Down syndrome, cystic fibrosis, Huntington's disease, sickle cell anemia, and muscular dystrophy.

Genetic disorders can be classified into different categories based on the type of genetic mutation involved. These categories include:

1. **Single-gene disorders:** These disorders are caused by a mutation in a single gene and can be further classified into autosomal dominant (e.g., Huntington's disease), autosomal recessive (e.g., cystic fibrosis), or X-linked (e.g., hemophilia) disorders, depending on the pattern of inheritance.

2. **Chromosomal disorders:** These disorders are caused by abnormalities in the structure or number of chromosomes. Examples include Down syndrome (trisomy 21), Turner syndrome (monosomy X), and Klinefelter syndrome (XXY).
3. **Multifactorial disorders:** These disorders result from a combination of genetic and environmental factors. They include conditions like heart disease, diabetes, and certain types of cancer. While genetics plays a role, other factors such as lifestyle, diet, and exposure to environmental toxins also contribute to the development of these disorders.
4. **Mitochondrial disorders:** These disorders are caused by mutations in the DNA of mitochondria, the energy-producing structures within cells. They can lead to a variety of health problems, including muscle weakness, neurological disorders, and organ dysfunction.

The field of medical genetics aims to understand the causes and mechanisms of genetic disorders, provide accurate diagnosis through genetic testing, offer genetic counseling to individuals and families at risk, and develop treatments or interventions to manage or prevent these disorders. Advances in genetic research and technology have significantly improved our ability to diagnose and treat genetic disorders, offering new possibilities for personalized medicine and precision therapies.

Forensic genetics and DNA profiling:

Forensic genetics, also known as forensic DNA analysis, is a branch of genetics that applies genetic techniques and principles to the field of forensic science. It involves the analysis and interpretation of DNA evidence for criminal investigations, identification of individuals, and paternity testing, among other applications.

DNA profiling, also referred to as DNA fingerprinting or genetic profiling, is a technique used in forensic genetics to determine the unique genetic makeup of an individual. It involves analyzing specific regions of an individual's DNA, which contain highly variable sequences known as polymorphisms or markers. These markers can be used to differentiate individuals and establish their identity.

The process of DNA profiling typically involves the following steps:

1. **Sample collection:** DNA samples can be obtained from various biological materials, such as blood, semen, saliva, hair, or tissue. The quality and quantity of the DNA sample are crucial for accurate analysis.
2. **DNA extraction:** The DNA is isolated from the collected sample using specialized laboratory techniques. This step aims to separate the DNA from other cellular components.
3. **Polymerase Chain Reaction (PCR):** PCR is a technique used to amplify specific regions of DNA. In DNA profiling, specific genetic markers, such as short tandem repeats (STRs), are targeted for amplification. These markers are highly variable among individuals.
4. **DNA analysis:** The amplified DNA fragments are analyzed using various methods, such as gel electrophoresis or capillary electrophoresis. These techniques separate the DNA fragments based on their size, allowing for the identification of the different alleles (variants) present at each marker.
5. **DNA database comparison:** The obtained DNA profile is compared to known DNA profiles stored in databases, such as national DNA databases or databases of convicted offenders. This step is crucial for identifying potential matches or suspects.
6. **Statistical analysis:** Statistical calculations are performed to determine the likelihood of a match between the DNA profile of the unknown sample and a known individual. This analysis takes into account the frequency of occurrence of specific alleles in the population.
7. **Interpretation and reporting:** The results of the DNA analysis are interpreted and reported in a forensic context. This includes providing information on the statistical significance of the DNA match or the likelihood of a relationship in paternity testing cases.

Forensic genetics and DNA profiling have revolutionized criminal investigations by providing robust scientific evidence for identifying suspects, linking individuals to crime scenes, and exonerating innocent individuals. They have become invaluable tools in the pursuit of justice and the establishment of biological relationships in various legal and forensic contexts.

Agricultural biotechnology and crop improvement:

Agricultural biotechnology refers to the application of scientific techniques and principles to improve agricultural productivity, sustainability, and the quality of crops. It involves the use of genetic engineering, molecular biology, and other advanced tools to manipulate the genetic material of plants, animals, and microorganisms.

Crop improvement is a specific area within agricultural biotechnology that focuses on enhancing the characteristics of crops to meet various agricultural challenges. Genetic studies play a crucial role in crop improvement by providing insights into the genetic makeup of plants and identifying genes responsible for specific traits.

Genetic studies in crop improvement involve several key steps:

1. **Genetic Mapping:** Scientists use various techniques, such as DNA sequencing and molecular markers, to identify and map the location of genes in the genome of a crop plant. This helps in understanding the genetic basis of important traits.

2. **Trait Identification:** Through genetic studies, researchers can associate specific genes or regions of the genome with desirable traits, such as disease resistance, increased yield, or improved nutritional content. This allows breeders to selectively breed plants with these traits.

3. **Gene Transfer:** Genetic studies enable the transfer of desirable genes from one organism to another, including unrelated species. This process, known as genetic engineering or transgenic technology, allows for the introduction of genes that confer beneficial traits into crop plants. For example, genes can be inserted to provide resistance against pests, diseases, or herbicides.

4. **Marker-Assisted Selection (MAS):** Genetic markers linked to desired traits can be used to accelerate the breeding process. By analyzing the presence or absence of these

markers, breeders can select plants with the desired traits more efficiently and accurately. This reduces the time and resources required for conventional breeding methods.

5. **Genome Editing:** Recent advancements in genetic studies have led to the development of genome editing techniques such as CRISPR-Cas9. This technology allows precise modifications to the plant's DNA, facilitating the targeted alteration of specific genes to enhance desirable traits or eliminate unwanted ones.

Overall, genetic studies in agricultural biotechnology and crop improvement provide valuable insights into the genetic basis of crop traits and offer tools to manipulate plant genomes for more efficient and targeted crop improvement strategies. These advancements contribute to the development of improved crop varieties with enhanced productivity, resistance to pests and diseases, and improved nutritional profiles, thus addressing the challenges of food security and sustainability in agriculture.

Conservation genetics and biodiversity preservation:

Conservation genetics is a field of study that combines principles from genetics, ecology, and conservation biology to understand and preserve the genetic diversity of populations and species. It focuses on how genetic factors influence the viability, adaptability, and long-term survival of populations in the face of various threats, such as habitat loss, climate change, and human activities.

Biodiversity preservation is a central goal of conservation genetics. Biodiversity refers to the variety of life on Earth, including the diversity of species, genes, and ecosystems. Preserving biodiversity is essential because it provides numerous ecological, economic, and cultural benefits. Genetic study plays a crucial role in biodiversity preservation by providing insights into the genetic makeup and evolutionary history of populations and species, which can inform conservation strategies and actions.

Here are some key aspects of conservation genetics and how they contribute to biodiversity preservation:

1. **Genetic Diversity Assessment:** Conservation geneticists examine the genetic variation within and among

populations. Genetic diversity is crucial because it enables populations to adapt to changing environmental conditions and enhances their resilience to threats. By analyzing DNA markers and genetic data, researchers can assess the genetic health and diversity of populations, identify unique or rare genetic variants, and determine the genetic structure of species across their geographic range.

2. **Population Viability Analysis:** Conservation geneticists use population genetics models and simulations to estimate the long-term viability of populations. They consider factors such as population size, gene flow, inbreeding, and genetic rescue to assess the risk of population decline or extinction. These analyses help determine effective population management strategies, such as translocations, captive breeding programs, or habitat restoration, to ensure the survival of endangered or threatened species.

3. **Conservation Prioritization:** Genetic data can inform conservation prioritization efforts by identifying populations or species with high genetic distinctiveness or unique evolutionary lineages. These genetically distinct populations are often prioritized for conservation actions as they may harbor important adaptive traits or represent evolutionarily significant units. Such information helps allocate resources and conservation efforts effectively.

4. **Conservation Breeding Programs:** Genetic studies provide insights into the mating systems, reproductive biology, and genetic compatibility of species. This information is valuable for managing captive breeding programs aimed at conserving endangered species. By avoiding inbreeding and maintaining genetic diversity in captive populations, conservationists can establish healthy and viable populations for future reintroduction into the wild.

5. **Monitoring and Adaptive Management:** Conservation genetics also plays a crucial role in monitoring the effectiveness of conservation efforts and guiding adaptive management strategies. By tracking changes in genetic

diversity over time, researchers can assess the impact of conservation interventions and make necessary adjustments to ensure long-term success.

Overall, conservation genetics contributes to biodiversity preservation by providing a scientific understanding of the genetic aspects of populations and species. By integrating genetic knowledge into conservation strategies, researchers and practitioners can make informed decisions to conserve and sustainably manage the Earth's biological diversity.

Chapter 12:
Practice Questions and Mock Tests

· **NEET-style questions and explanations**
· **Mock tests to assess understanding and preparedness**

NEET-style questions and explanations

Question: In a population of fruit flies, a gene called "wingless" (W) exhibits incomplete dominance. The genotype WW produces flies with fully developed wings, genotype Ww produces flies with partially developed wings, and genotype ww produces flies with no wings. In a random mating population of 1000 fruit flies, the genotype frequencies for WW, Ww, and ww were found to be 0.49, 0.42, and 0.09, respectively.

Calculate the allele frequencies for the "wingless" gene in the population.

Explanation: To calculate the allele frequencies, we need to consider the genotype frequencies and the assumption of Hardy-Weinberg equilibrium.

Let's assume p represents the frequency of allele W, and q represents the frequency of allele w in the population.

According to the Hardy-Weinberg equilibrium equation, $p^2 + 2pq + q^2 = 1$, where p^2 represents the frequency of genotype WW, 2pq represents the frequency of genotype Ww, and q^2 represents the frequency of genotype ww.

Given that the genotype frequencies for WW, Ww, and ww are 0.49, 0.42, and 0.09, respectively, we can set up the following equations:

$p^2 = 0.49$ -- (1) $2pq = 0.42$ -- (2) $q^2 = 0.09$ -- (3)

From equation (1), we can take the square root of both sides: $p = \sqrt{0.49} = 0.7$

Now, substituting the value of p in equation (2), we can solve for q: $2(0.7)q = 0.42$ $1.4q = 0.42$ $q = 0.42/1.4$ $q = 0.3$

Therefore, the allele frequencies for the "wingless" gene in the population are $p = 0.7$ and $q = 0.3$.

Note: In this example, we assumed that the population is in Hardy-Weinberg equilibrium, which means that the population is large, random mating is occurring, there is no migration, mutation, or natural selection.

Question: Which of the following inheritance patterns is characterized by a trait that appears in every generation, affecting both males and females equally, and is transmitted through both male and female parents?

a) Autosomal recessive inheritance b) Autosomal dominant inheritance c) X-linked recessive inheritance d) X-linked dominant inheritance

Explanation: The correct answer is b) Autosomal dominant inheritance.

Autosomal dominant inheritance is characterized by the following features:

1. Trait appears in every generation: In autosomal dominant inheritance, affected individuals are present in every generation of the pedigree. This is because the trait is caused by a dominant allele, and individuals carrying at least one copy of the allele will express the trait.

2. Affects both males and females equally: Autosomal dominant traits do not show a sex bias. Both males and females can be affected by the trait in equal proportions.

3. Transmitted through both male and female parents: In autosomal dominant inheritance, an affected individual can transmit the trait to their offspring regardless of their gender. The trait can be passed from either an affected mother or an affected father to their children.

On the other hand, let's briefly explain the other options:

a) Autosomal recessive inheritance: In autosomal recessive inheritance, the trait appears in a recessive manner, meaning an individual must inherit two copies of the recessive allele (one from each parent) to express the trait. Typically, the trait skips generations and affects males and females equally.

c) X-linked recessive inheritance: X-linked recessive inheritance involves a gene located on the X chromosome. The trait is more commonly expressed in males because they have only one X chromosome, while females have two. Males inherit the X chromosome from their mother, so if the mother is a carrier or affected, the sons have a higher chance of being affected. Daughters can be carriers if their father is affected.

d) X-linked dominant inheritance: X-linked dominant inheritance occurs when a gene is located on the X chromosome, and the trait is expressed when at least one copy of the dominant allele is present. The trait can affect both males and females, but it often shows a higher frequency in females. Affected males always pass the trait to all of their daughters but not to their sons.

Question: Which of the following genetic disorders is caused by a mutation in a single gene? a) Down syndrome b) Hemophilia c) Turner syndrome d) Cystic fibrosis

Explanation: The correct answer is d) Cystic fibrosis. Cystic fibrosis is caused by a mutation in a single gene called the cystic fibrosis transmembrane conductance regulator (CFTR) gene. This mutation affects the production and function of a protein involved in the regulation of salt and water movement in cells, leading to the buildup of thick, sticky mucus in various organs.

Question : Which of the following genetic techniques is used to determine the order of nucleotide bases in a DNA molecule? a)

Polymerase chain reaction (PCR) b) Gel electrophoresis c) DNA sequencing d) Restriction enzyme digestion

Explanation: The correct answer is c) DNA sequencing. DNA sequencing is a technique used to determine the precise order of nucleotide bases (adenine, cytosine, guanine, and thymine) in a DNA molecule. It allows researchers to obtain the genetic code of an organism and identify specific genetic variations or mutations.

Question: What is the function of transfer RNA (tRNA) in protein synthesis? a) Transcription of DNA into RNA b) Translation of RNA into proteins c) Binding of amino acids to ribosomes d) Carrying genetic information in the form of codons

Explanation: The correct answer is c) Binding of amino acids to ribosomes. Transfer RNA (tRNA) is responsible for carrying amino acids to the ribosomes during protein synthesis. Each tRNA molecule recognizes a specific codon on the messenger RNA (mRNA) molecule and brings the corresponding amino acid, which is then incorporated into the growing polypeptide chain.

Question: Which of the following is an example of a sex-linked genetic disorder? a) Huntington's disease b) Duchenne muscular dystrophy c) Tay-Sachs disease d) Marfan syndrome

Explanation: The correct answer is b) Duchenne muscular dystrophy. Duchenne muscular dystrophy is an example of a sex-linked genetic disorder. It is caused by a mutation in the dystrophin gene located on the X chromosome. Since males have one X and one Y chromosome, a single mutated copy of the dystrophin gene is sufficient to cause the disorder, whereas females need two mutated copies.

Question: In a DNA molecule, adenine (A) always pairs with: a) Thymine (T) b) Guanine (G) c) Cytosine (C) d) Uracil (U)

Explanation: The correct answer is a) Thymine (T). In DNA, adenine (A) always pairs with thymine (T) through hydrogen bonding. Similarly, guanine (G) always pairs with cytosine (C). This base-pairing pattern is essential for the stability and replication of DNA molecules.

Question: Which of the following genetic disorders is caused by a deletion of a portion of chromosome 15? a) Down syndrome b)

Duchenne muscular dystrophy c) Cri-du-chat syndrome d) Turner syndrome

Explanation: The correct answer is c) Cri-du-chat syndrome. Cri-du-chat syndrome is a rare genetic disorder caused by the deletion of a portion of chromosome 15. Individuals with this syndrome typically have characteristic high-pitched cries resembling the mewing of a cat, intellectual disability, and developmental delays.

Question: A couple has a child with cystic fibrosis, an autosomal recessive disorder. The father is unaffected, but the mother is a carrier of the cystic fibrosis gene. What is the probability that their next child will have cystic fibrosis? a) 25% b) 50% c) 75% d) 100%

Explanation: The correct answer is b) 50%. Cystic fibrosis is an autosomal recessive disorder, which means that both copies of the CFTR gene must be mutated for an individual to have the disease. In this case, the father does not have the disease and must have two normal copies of the gene. The mother is a carrier, meaning she has one normal copy and one mutated copy. When they have a child, there is a 25% chance that the child will receive the mutated gene from both parents and have cystic fibrosis. Therefore, the probability is 50%.

Question: In a population of butterflies, the color of their wings is determined by a single gene with two alleles, B and b. If 36% of the butterflies have the genotype BB and 16% have the genotype bb, what is the frequency of the B allele in the population? a) 0.36 b) 0.64 c) 0.72 d) 0.84

Explanation: To calculate the frequency of the B allele, we need to determine the proportion of the population that carries at least one copy of the B allele. The frequency of the B allele can be calculated as the square root of the proportion of butterflies with the BB genotype plus half the proportion with the Bb genotype.

Frequency of B allele = $\sqrt{}$(Proportion of BB genotype + 0.5 * Proportion of Bb genotype) = $\sqrt{}$(0.36 + 0.5 * 0) = $\sqrt{0.36}$ = 0.6 . Therefore, the frequency of the B allele in the population is 0.6 or 60%.

I hope these questions and explanations help you in your genetic studies!

Mock tests to assess understanding and preparedness:

Mock tests are a valuable tool for assessing understanding and preparedness in genetic studies. They help identify knowledge gaps, highlight areas that require further study, and provide a sense of the exam format and time management skills. Here are a few sample mock test questions in the field of genetic studies:

1. True or False: Genetic information is stored in the form of a chemical code called DNA.
2. Which of the following is responsible for the transmission of genetic information from one generation to the next? a) Messenger RNA (mRNA) b) Transfer RNA (tRNA) c) Ribosomal RNA (rRNA) d) Deoxyribonucleic acid (DNA)
3. Which process involves the creation of a complementary RNA molecule based on a DNA template? a) Transcription b) Translation c) Replication d) Recombination
4. The complete set of genetic material present in an organism is called its: a) Genome b) Genotype c) Phenotype d) Allele
5. In a monohybrid cross between two heterozygous individuals (Aa x Aa), what is the expected phenotypic ratio of the offspring? a) 1:1 b) 1:2:1 c) 3:1 d) 9:3:3:1
6. Which genetic disorder is caused by the presence of an extra chromosome 21? a) Down syndrome b) Turner syndrome c) Klinefelter syndrome d) Cystic fibrosis
7. What is the function of restriction enzymes in genetic engineering? a) They cut DNA at specific recognition sequences. b) They synthesize new DNA strands. c) They amplify DNA fragments. d) They insert DNA into host cells.
8. What is the purpose of gel electrophoresis in genetic analysis? a) To visualize DNA under a microscope. b) To separate DNA fragments based on size. c) To introduce foreign DNA into cells. d) To determine the DNA sequence.
9. Which technique allows the amplification of a specific DNA sequence in large quantities? a) Polymerase chain reaction (PCR) b) Southern blotting c) DNA sequencing d) Fluorescence in situ hybridization (FISH)

10. What is the term for a change in a DNA sequence that leads to the production of an altered protein? a) Mutation b) Polymorphism c) Hybridization d) Translocation

Appendix: Glossary of Genetic Terms

Here's a glossary of genetic terms for your reference:

1. Allele: Alternative forms of a gene that occupy the same position (locus) on a chromosome.
2. Chromosome: A thread-like structure made up of DNA and proteins found in the nucleus of cells. Chromosomes carry genetic information.
3. DNA: Deoxyribonucleic acid, a molecule that contains the genetic instructions used in the development and functioning of all known living organisms.
4. Gene: A segment of DNA that contains the instructions for building a specific protein or carrying out a specific function.
5. Genotype: The genetic makeup of an individual, referring to the combination of alleles present for a particular trait.
6. Phenotype: The observable characteristics or traits of an organism that result from the interaction between its genotype and the environment.
7. Mutation: A change in the DNA sequence of a gene or chromosome. Mutations can be inherited or occur spontaneously.
8. Genome: The complete set of genetic material (DNA) present in an organism.
9. Genomics: The study of an organism's entire genome, including the interactions between genes and their environment.
10. Genetic variation: The diversity of genetic material within a population or species. It arises through mutations, recombination, and genetic drift.

11. Homozygous: Having two identical alleles for a particular gene.
12. Heterozygous: Having two different alleles for a particular gene.
13. Dominant: An allele that is expressed when present in either the heterozygous or homozygous state.
14. Recessive: An allele that is expressed only when present in the homozygous state.
15. Genetic disorder: A condition caused by abnormalities or mutations in genes or chromosomes that affect the normal functioning of the body.
16. Genetic engineering: The process of manipulating an organism's genes to introduce new traits or modify existing ones.
17. Genotype-phenotype correlation: The relationship between an individual's genetic makeup (genotype) and the observable traits (phenotype) that result from it.
18. Genetic counseling: The process of providing information and support to individuals or families about genetic conditions, inheritance patterns, and the risks of passing on genetic disorders.
19. Genomic medicine: The use of genetic information, including genomics and personalized genomics, in medical research, diagnosis, and treatment.
20. Genetic screening: The process of testing individuals for the presence of specific genetic mutations or variations that may be associated with certain diseases or conditions.

Please note that this glossary provides basic definitions, and some terms may have more complex or specific meanings in different contexts or fields of genetics.

"Genetics Simplified" aims to provide a comprehensive overview of genetics, specifically tailored for NEET (National Eligibility cum Entrance Test) examination preparation. The book covers all essential topics within the genetics domain, presenting them in a simplified manner with clear explanations and diagrams. Each chapter concludes with practice questions and mock tests designed to reinforce understanding and help students assess their knowledge.

By studying this book, aspiring medical and biology students will gain a solid foundation in genetics, ensuring they are well-prepared to tackle the genetics section of the NEET examination.